Advanced Excel for Surveyors

Advanced Excel for Surveyors

Philip Bowcock
and
Natalie Bayfield

2003

ESTATES GAZETTE
151 WARDOUR STREET, LONDON W1F 8BN

First published 2003

ISBN 0 7282 0413 4

Typeset by Amy Boyle, Rochester, Kent
Printed and bound by The Cromwell Press, Trowbridge, Wiltshire

Contents

Foreword

My colleagues fall into three fairly distinct categories, generally divided by age: those for whom computers are dangerous and untrustworthy, those for whom computers are life blood, and a middle category (in which I count myself) who have trained themselves to use computers but do not have the instinctive empathy with information technology possessed by the younger generation.

This work will not convert the non-believers, but has considerable value for the other two categories. The new generation of surveyors will eagerly absorb the extra features and dimensions highlighted in these pages, including workbook sharing and the numerous options for making life easier through shortcuts. Mysterious areas like macros and regression analysis are also opened up in a logical and comprehensible way.

The converted generation will, as I did, understand the sheer scale of the possibilities which Excel offers, and implementing some of these will extend their computer literacy. By way of example, as a fund manager I am obliged to be interested in the correlation between different investment classes. These pages reveal how simple it is to calculate correlation coefficients and demonstrate where Excel can play its part in the crucial area of risk analysis.

We are told that in our lifetime we use only a fraction of the capacity of our brains. It is patently clear from this book, and its predecessor, that the same applies to Excel. You will find that the following pages open many new doors.

DAVID HUNTER
Chief Executive
Aberdeen Property Investors UK Limited

x

Preface

Philip Bowcock was formerly Lecturer in Valuation at the University of Reading until his retirement after a total of 29 years in the Department of Land Management and prior to that in the College of Estate Management before the merger with the University in 1972. He has always been interested in mathematics and has been involved in the use of computers for the past 30 years.

Natalie Bayfield is director of Bayfield Training Limited an established Financial Modelling Consultancy and Training Organisation. She lectures regularly at City and Southbank Universities. She is also course director for the Investment Property Forum's popular 'Property Applied Spreadsheets and Systems' (PASS) course, as well as course director for Euromoney's 'Modelling Techniques for Structuring Real-Estate Finance'.

The formal name of the software discussed is Microsoft Excel™, but for convenience we shall refer to this simply as 'Excel'. Screen shots are reproduced by kind permission of The Microsoft Corporation.

This volume is intended to extend the discussion of Excel in our previous volume, *Excel for Surveyors*, to more advanced problems relevant to the work of surveyors and property advisors. It does not set out or recommend any particular format or method for making decisions.

The text was produced on a Toshiba PC using Microsoft Windows and Microsoft Office 2000 (Natalie) and a Macintosh G4 933 using Microsoft Office 2001 (Philip) with final editing using Microsoft Office XP. We found few practical differences between these versions and the previous versions of Excel (Office 97 and Office 98 respectively). Final screen shots were edited where necessary with Adobe Photoshop 7.0™, except those specifically related to the Macintosh.

Much of the content of this volume can be used with earlier versions of Excel, and it should be possible to set up most of the earlier examples with no modification.

We are very grateful to the staff of *Estates Gazette* and to Colin Greasby and our editor, Audrey Andersson, in particular, for their encouragement in proceeding with this work.

Philip Bowcock and Natalie Bayfield

Chapter 1

Introduction

Pythagoras observed that the Egyptians and Babylonians conducted each calculation in the form of a recipe which could be followed blindly. The recipes, which would have been passed down through the generations, always gave the correct answer and so nobody bothered to question them or explore the logic underlying the equations. What was important for these civilizations was that a calculation worked – why it worked was irrelevant.

Simon Singh

One might boldly suggest that in the property world bespoke valuation software could be considered a modern day analogy of these ancient 'recipes'. The importance of bespoke software such as packages sold by Circle, Systemslink3 and Kell to name only a few developers is in the ease and consistency that they confer to users and their firms. The prevalence of these packages among property firms suggests this to be the case. It also suggests that there is a broad acceptance that they work. We have not examined any one package in detail, but it is safe to assume that these packages are tested many times by more than one person. By comparison, this is more that can be said for most appraisal models created by surveyors on Excel and others with wider responsibilities[1].

What is the point in trying to replicate models in Excel? There are two fundamental reasons why it is important to know why the models work. First, due diligence – the valuer might need to justify a particular approach to the problem. Second, without knowing how the models work it is easy to generate an error when adjusting for non-standard circumstances[2]. However forward thinking software houses might be, their packages are still machines producing answers to questions posed by surveyors and not by software developers. There will always be a lead time in creating these machines, unless, of course, the surveyor can create the machine himself. Fortunately surveyors do not have to be software developers. A working knowledge of Excel will normally do. However they will need to revise their valuation skills. We are concerned that many new recruits into surveying continue to be taught to use tables and valuation packages, without appreciating that constructing models from scratch is the most reliable way to build appraisal skills.

One other undeniable feature of Excel is its compatibility. It is becoming fashionable to communicate appraisals electronically instead of, or at least as well as, delivering the printed version. Some of the larger firms also build in some interactivity for their audience allowing clients to change yields and bank rates and watch the changes ripple through the model to the result. It is only the largest surveying firms that have all of the valuation packages, and can read any

---

valuation in any medium. By setting up a model in Excel one can be confident that just about any firm or individual can be persuaded to accept it. Perhaps most importantly this is also true for the new generation of clients in funds and institutions who are less likely to have the bespoke programs.

We have recognized that worksheet skills in surveying are more widespread, and often quite complex. It is a marked contrast with only a few years ago when many surveyors did not even have a computer sitting on their desk. However, as we have mentioned, with increased complexity comes increased possibility for error. We do not recommend that you build an 'all singing dancing' model. Rather, use Excel to deal with a specific problem. Remember that it is safe practice to seek a second opinion on your work before passing it on to a client.

Multi-tenanted properties present some of the more difficult challenges to the construction of worksheets. What makes them especially difficult is the frequently irregular nature of the individual tenant cash-flows, unexpired terms, varying review patterns, rents and growth rates, and miscellaneous costs. These all make the search for a one-size-fits-all formula virtually impossible. The many different tools available in Excel to assist in this task means that there are almost as many different models as there are surveyors making them. Some surveyors pride themselves on the complexity of their worksheets. However the most complex are not necessarily the best. Most Excel textbooks discuss the merits of planning for a 'good' worksheet, and this book is no exception. There are some simple rules and guidelines that can be applied at the planning stage to make your worksheets simpler and easier to understand while doing the same job.

The most unreliable worksheets are templates that have been set up by one surveyor and then added to over the years by others. These worksheets often gain credence from longevity, but may contain subtle flaws. Any worksheet that earns itself the honour of being called a template must be offered to those who use it only with training or with a fully detailed labelling and description. Any user is entitled to question its structure.

There are some formatting tips which may be employed to help surveyors read and understand worksheets and templates quickly. We will discuss these later, and in addition we shall introduce some other Excel functions that can cut down on the technical complexity of the worksheet.[1]

Although much of the material was initially produced on a Macintosh computer all examples and references are based on PC operations. We assume that Macintosh users will be familiar with the minor variations in procedures. Excel and other Office files are interchangeable between the two operating systems.

---

[1]   By way of example, one of the authors was once presented with calculations of the value of a development project done first by a proprietary package and secondly by Excel worksheet. The answers were different, but the reason for this could not be found.

Chapter 2

Some of the new features in Excel

2.1 Introduction

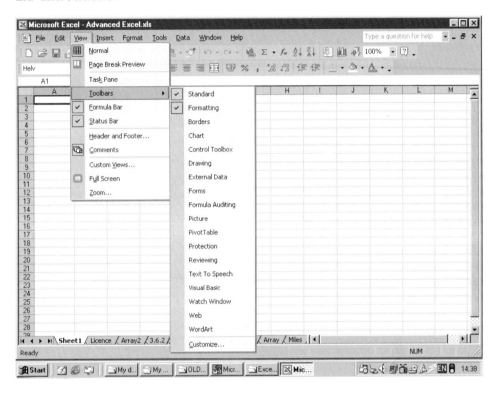

In this section we set out some of the more useful modifications and additions to the latest versions of Excel. There are now several versions current in the different issues of Microsoft Office – Office 97, Office 2000, Office ME and Office XP, plus Macintosh versions 1998, 2001 and 10.0 – and the following facilities are not found in all versions. Some facilities are just re-arrangements of menu items, but others are significant and useful additions. The general operation of Excel is as before – a tool which can be used by surveyors to perform operations which would be too time-consuming, difficult, or impossible by other means.

You should appreciate that Excel has many hundreds of features and facilities, and they are not all considered here – it would require a very large volume to cover every possibility, and many of these would not be relevant to the work of surveyors. For example they include several which involve imaginary numbers for which, so far, we are unaware of any purpose in surveying. Similarly, many of the financial functions are based on American practice which differs from that in this country.

If you upgrade to the latest version you can still open and save all your previous work in Version 97/98 format. You can even save in earlier versions right back to Version 2.2, although some formatting and other functions may be lost.

2.2 New features

2.2.1 *Help*

You should bear in mind that Excel is a very elaborate program with many facilities including many specialised operations. Although we discuss many of those which are of particular use to surveyors, there are inevitably other aspects and methods which may be useful in particular cases. The on-screen **Help Menu** ⇒ **Microsoft Excel Help** (or Function Key F1) is a valuable facility which is well worth exploring.

2.2.2 *FILE MENU⇒PROPERTIES*

This produces a window with five tabs of which 'General', 'Statistics' and 'Contents' give general information about the workbook. 'Summary' and 'Custom' provide locations where further information about the workbook, its contributors and destination can be saved.

2.2.3 *More toolbars*

Several new toolbars are now available as standard, of which the Pivot Table, Visual Basic and Web are possibly the most useful. As with other toolbars, the purpose is simply to make the relevant buttons conveniently available, and the more you open the less space there will be to see the worksheet. Call them and close them as necessary.

2.2.4 *Custom views*

You can create a custom view to define sets of particular display and print settings and save them as views with **View Menu⇒Custom Views**. You can switch to any of the views whenever you want to display or print the workbook in a different way. The stored settings include column widths, display options, window size and position on the screen, window splits or frozen panes, the sheet that is active, and the cells that are selected at the time the view is created. There is also the option to save hidden rows, hidden columns, filter settings, and print settings.

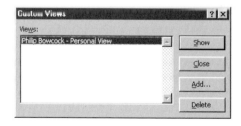

The custom view includes the entire workbook. If you hide a sheet before a view is added, Microsoft Excel hides the sheet each time you show that view.

Before you create a view set up the workbook as you want to view and print it. If you include print settings in a view, the view includes the defined print area or the entire worksheet if the sheet has no defined print area.

2.2.5 Number formats – fractions, phone numbers, etc

Several new number formats are now included, of which Fractions and Phone Numbers would appear to be the most useful in this country. (Others such as Zip codes and social security numbers are based on American practice.)

(a) If a cell is defined as a fraction any fractional number will be converted to the form selected in the **Custom** type box. Fractions are defined using the '?' character with up to three denominator digits, thus ???/??? will convert 0.123456 to 10/81. The number is of necessity rounded to the nearest fraction but is still preserved in decimal form for calculation purposes.

Numbers can also be entered as fractions and will be converted to decimal form internally. Entering a quarter as a fraction will convert to 0.25. Note however that if the cell is not formatted as a fraction this would be interpreted as a date – 4th January!

(b) To format a 7-digit telephone number go to **Format Menu⇒Cells** and under the NUMBER tab select CUSTOM. In the TYPE box insert

####

Note the space. STD codes can be included with a third set of hash characters, but obvious problems arise with codes in areas such as London.

You should appreciate that both fractions and telephone codes are still stored as numbers (as is also the case with dates etc) and these can be sorted and also be the subject of calculations (if that ever serves a useful purpose so far as telephone codes are concerned).

2.2.6 WYSIWYG

The Font menu is now WYSIWYG ('What you see is what you get'). Font names now display in their font on the Font menu so you can preview a font style before you apply it to your worksheet.

2.2.7 Colour Picker

The Colour Picker allows you to select a colour for your drawings which does not appear among the 40 standard colours on the Drawing toolbar buttons.

2.2.8 Euro symbol

There is now a Euro currency symbol and additional number formats are available with this.

2.2.9 Shared tools

Consistency in shared tools across Office programs means that you only need to learn one way to get your work done.

2.2.10 Border Drawing

The new Border Drawing toolbar enables you to define and draw your own borders, selectively erase borders, merge borders and cells in a single click, and change border styles and colours.

2.2.11 Chart menu

New facilities in the Chart menu give increased capacity, and high-end display options give you powerful ways to analyze your data visually.

If your chart values consist of large numbers, you can make the axis text shorter and more readable by changing the display unit of the axis. For example, if the chart values range from 1,000,000 to 50,000,000, you can display the numbers as 1 to 50 on the axis and show a label to indicate that the units express millions.

Multiple-level category labels When the category data on your worksheet is hierarchical (that is, consists of more than one heading level), Excel automatically maintains this hierarchy in your chart by including each level in the category axis labels.

2.2.12 Creating formulas

Creating formulas on worksheets is simplified, and new tools help you build and edit formulas and enter functions.

Clicking on the AutoSum button now gives you the options of Average, Count, Max, and Min, making it easier to create formulas by using these functions.

2.2.13 Web Queries

Web Queries have been improved. You can create and run queries to retrieve data that is available on the World Wide Web. You can either select an entire Web page or specify a table on a Web page to retrieve. There are several sample Web queries that you can run. Data from text files can now be brought into Excel in the same way that you create other database queries.

2.2.14 File conversions

It is possible to convert a FileMaker file into a Macintosh Worksheet and conversely.

2.2.15 Euroconvert

The **Euroconvert** function will convert pounds sterling to euros and vice versa It will also convert between currencies of the European Union, but since most of these have been absorbed into the Euro system this facility is largely redundant.

2.2.16 Pivot Tables

The Layout reports of Pivot Tables appear directly on worksheets. After you click Finish in the PivotTable Report Wizard, blue outlined drop areas appear on your worksheet, and the PivotTable toolbar displays a list of the fields from your source data. You can lay out the PivotTable report directly on the worksheet by dragging the fields from the toolbar to the drop areas. You no longer have to use PivotTable selection when you format a PivotTable report. Formatting applied by using regular Excel selection is retained when you refresh or change the layout.

Chapter 3

Managing large worksheet models and good practice

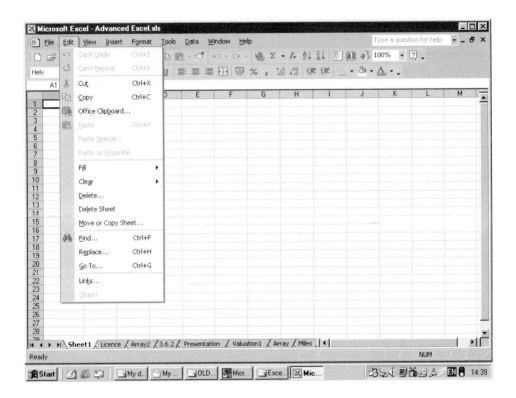

3.1 Navigating large worksheets

Examples of large worksheets in property include databases, multi-tenanted cash flows, portfolio analysis models and some more complex statistical analysis models.

With any large set of data there will be navigational requirements eg the ability to locate a particular value quickly, jumping to the end of a row or a column, or keeping data labels visible while scrolling through the dataset. All these functions serve to make large worksheets more manageable. They also represent one area where keyboard shortcuts are usually the most expedient method.

3.1.1 Splitting the screen

The problem with using scrollbars to move around is that they move data labels out of the way as well. The way around this is to split the screen into two. This was explained in *Excel for Surveyors* as selecting **Window Menu**⇒**Split Screen**. In fact these screen splits are permanently available on the Excel screen.

(a) Make sure you have cell A1 visible.
(b) Rest the mouse pointer over the grey bar to the right of the bottom scrollbar until you see a double headed arrow.
(c) Drag this across the worksheet until the line divides the screen into two convenient sections. Alternatively you can double click on the arrow. The screen will be split at the selected cell and the scroll bar at the bottom of the screen will divide into two sections to match. Click anywhere in a section of the screen to make that section active. You can then use the scroll bar for that section to move through it, while the other section can be kept in view.

Similarly, the screen can be split horizontally in an equivalent position above the vertical scrollbar can be used to split the screen horizontally by dragging or clicking on the small grey bar immediately above the vertical scroll bar. Finally you can use both operations to divide the screen into four sections.

To restore the screen to normal drag the split line to one end of its scroll bar.

3.1.2 A useful shortcut

When inputting a large amount of data you can use shortcut keys for copying information rather than going back to the mouse. In the following example of a simple license fee schedule in which cells C2:C6 are to hold the same formula :

	A	B	C	D
	Unit	Licence Fee	VAT	Total
1		per month		
2	1	660	1.175	775.50
3	2	606	1.175	712.05
4	3	850	1.175	998.75
5	4	430	1.175	505.25
6	5	606	1.175	712.05
7				

(a) Highlight cell C2.
(b) Use the keyboard strokes, Ctrl & C, to copy.
(c) Hold down the Shift key and press the Down arrow four times to reach the end of the column, or drag over cells C3:C6.
(d) Use the keyboard strokes, Ctrl & V, to fill the selected cells with the copied formula.

3.1.3 The mouse shortcut

An alternative little trick can do the same job, particularly if your column of data extends far beyond the bottom of your screen.

(a) Select cell D2.

(b) Hover on the fill handle in the bottom right hand corner of the selection as if you were about to copy.

(c) Double click. This should fill the remaining cells with the formula copied from the first cell.

3.2 Formatting and the AutoFormat command

3.2.1 *Alignment*

The object of formatting is to make particular cells stand out to distinguish between different types of information in a worksheet. To do this we can apply borders to cells or shade cells with a background colour or colour pattern.

The header of a column is often much wider than the data in that column. An alternative to creating unnecessarily wide columns or abbreviated labels is to rotate the text of the header and apply borders that are rotated to the same degree as the text. Select the header row and **Format Menu**⇒**Cells**, click the **Alignment** tab, and use the orientation lever to rotate the text to the desired angle.

3.2.2 *Styles*

We discussed simple examples of the formatting of cells in *Excel for Surveyors* but there are many more options available. A particularly useful facility is the ability to create styles in a manner similar to that of Microsoft Word. Go to **Format Menu** ⇒**Style** to open this box.

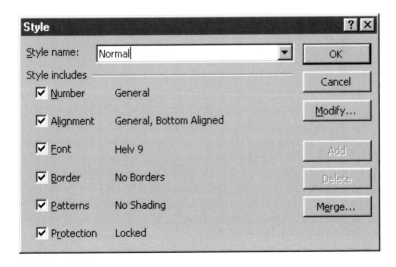

The default style is Normal, and this will be the default for all the worksheets in the workbook. The Normal style can be changed with the Modify button, and doing this will be reflected immediately throughout the workbook. A new style can be created by changing the style name and specifying the new characteristics with the **Modify** button which takes you to the **Format Cells** dialogue box. This will then be added to the list of styles. Thereafter, selecting a cell or a range and selecting the list with the **Format Menu⇒Style** ... command will change the cell format to that specified. Styles available include currency, percentages, and commas to separate thousands.

Caution – Change the style name first: if you change the characteristics without changing the **Normal** style name you will alter this throughout the worksheet.

If you have already formatted some cells on a worksheet to your own specification, you can use **Format Painter** to copy the formatting to other cells – select the cell with the desired format, click the **Format Painter** button on the standard toolbar, and then select the cells to which you want to apply the format.

The **Format Menu⇒Autoformat** ... command provides several useful formats which can be applied either to an entire list or to another large range that has distinct elements – for example, column and row labels, summary totals, and detail data. The design uses distinctive formats for the various elements in the range. For example List 1 will format alternate rows to a contrasting colour for easier reading. If you enter rows at the end of a list that you have already formatted, the formatting will be automatically extended if you have turned automatic formatting on.

3.3 Appropriate number formatting

Where possible you will want to avoid clutter and a good choice of number and text formatting and colour can make a major contribution to the appearance of the worksheet.

The Accounting and Currency formats are particularly useful here, as they allow for dozens of international currencies. The most useful ones for our purposes are obviously the Pound, Dollar and Euro signs.

Decimal places to the right of a number, even if zero, convey the accuracy to which the calculation is done. This is especially important in Excel which defaults to an accuracy of about 16 significant figures. In many calculations of course this is an advantage. However it is possible for errors to arise when working with some kinds of numbers, particularly currency. What we need to convey instead is the 'accuracy that matters'.

Excel will normally display a number to the nearest decimal place set as a format[1] even though it still stores and uses the most accurate number that it can calculate. A yield may appear in Excel as 8% while the actual result is 7.5%. If the result happened to be 7.65% we would still be interested in this precision. We may be less concerned that the yield was actually 7.648%, but the difference between this figure and 8% is an unacceptable assumption. We can however override the precision of Excel with the **Round, Roundup** and **Rounddown** functions. These are particularly useful when dealing with accounts as they prevent small differences accumulating to display an apparent error.

---

[1] It will round up if the last decimal place is greater than 5 and round down if it is 4 or under. For example 7.64 will be rounded to 7.6, whereas 7.65 will be rounded to 7.7.

3.3.1 Accounting format

The Accounting number format without currency symbols is normally the neatest for large currency data sets. The format lines up all the numbers slightly away from the right hand side of the cell to allow for the standard accounting practice of displaying parenthesis '( )' to indicate negative numbers. Commas are automatically placed at every group of three digits, and two decimal places to the right indicate pence. This format can be selected by clicking on the 'comma style' button on the formatting tool bar.

3.3.2 Currency format

This format allows the £ or relevant currency symbol to be placed with the label eg Rent £. and displays zero figures simply as a single dash. This lack of clutter gives a much crisper image to cash flows. It is often helpful to use currency format for totals while leaving intermediate amounts in normal format. If the same currency symbol appears in every cell this can increase rather than decrease clutter.

3.3.3 Date format

Dates of course are important in cash flows. It is easier on the eye to use the name of the month instead of its number as this naturally distinguishes the date from the rest of numbers in the DCF. The shortened month name is a preferable format, since this avoids the problem of unequal column or row sizes, as caused by May and September for example.

3.3.4 Custom format

We gave examples of custom formats in *Excel for Surveyors*, para 4.7. These allow you to include text in a number to describe what that number is. For example, formatting a number as 'Present Value of £1,' ## 'years' will create a number format in one cell which, for calculation purposes is a number but visually explains what that number is.

Note that, since it is a number, it cannot be word-wrapped, and therefore the cell must be wide enough to accommodate it, otherwise it will be filled with hash symbols. The solution to this problem is to double click on the cell division mark at the top of the worksheet and the column will be widened to accommodate the data.

3.4 Setting out

3.4.1 Presentation

The overall presentation is just as important as localised formatting and correctness. You will want to create an overall image that looks professional, but also allows the information contained in the worksheet to be quickly understood. Simplicity can convey professionalism, and a logical layout based on ordering information according to what the reader will want to see first will make the worksheet easy to understand.

For this reason most bespoke program's result pages show the inputs and outputs of a DCF together, leaving the actual cash flow until last. This is because the cash flow itself is relatively unimportant. It is a means to an end. What analysts and investors are really interested in is: What happens to the outputs if I change some or all of the inputs? For example: what happens to the NPV if I change the growth rate? This is information that they will want to see first. Borders and shading can help to distinguish between inputs and outputs.

3.4.2 Guiding the eye.

If there are several sheets in a workbook, or your worksheet will print several pages, eg multi-tenanted investment cash flows, it is advantageous to make the sheets 'look' as similar as possible. Doing so allows readers to 'learn' a layout from the first page/sheet and apply this knowledge to the next. Auction catalogues are a good example of this. Yields, prices, tenures, address are all put in the same place for each property on each page of the catalogue. This is a less obvious concept when it comes to worksheets, but the advantage is possibly greater since the information contained is likely to be more complex, and any extra speed gained will make the job of understanding the model less stressful.

The **bold** button on the formatting tool-bar is the simplest and most obvious way of picking out information that you wish to draw attention to, for example Totals, Net Present Value, or 'Profit or Loss'. Other methods include underline or increasing the size of the font. It is not just final figures, however, that can benefit the worksheet from being given a prominence. Cash Flows on first glance present themselves as a vast unintelligible sea of numbers. Their meaning can be gleaned from reading the column and row titles, but again this takes time. Conditional formatting is an excellent automatic way of highlighting rental figures in cells that represent a reviewed rent. By setting a condition that changes the number in the cash flow to a bold format if it is different to the figure above or the left (depending on the orientation of the cash flow) all rents due for renewal will be prominent.

Colour cells which are important, or which have some characteristic in common, using the Fill Colour button. There are 40 different colours to choose from, but of course too many can make the worksheet appear more confusing rather than less.

3.4.3 Example

In the following worksheet all rents reviews are emboldened.

	A	B	C	D	E	F	G	H	I
1	Property	24 Mar 01	24 Mar 02	24 Mar 03	24 Mar 04	24 Mar 05	24 Mar 06	24 Mar 07	24 Mar 08
2	1	100,000	100,000	100,000	**120,000**	120,000	120,000	120,000	120,000
3	2	80,000	**110,000**	110,000	110,000	110,000	110,000	**125,000**	125,000
4	3	150,000	150,000	150,000	150,000	150,000	**175,000**	175,000	175,000
5	4	140,000	140,000	140,000	140,000	140,000	**165,000**	165,000	165,000
6	5	95,000	95,000	**120,000**	120,000	120,000	120,000	120,000	**145,000**
7	6	50,000	50,000	50,000	**70,000**	70,000	70,000	70,000	70,000
8									

To achieve this result proceed as follows:

(a) Select cell C2.
(b) Select **Format Menu⇒Conditional Formatting**.
(c) Select 'not equal to' from the drop down menu in the second. (There is an option to 'Add more conditions' but in this example we want to format for just one, so click OK again.)
(d) In the third text box type the reference for the first rent, ie B2. Caution – enter the cell reference manually as a relative reference – clicking on a cell (B2 in this case) may enter the absolute reference of $B$2 which will not give the desired result.
(e) Click the Format button.
(f) Click on Bold under font style and click OK. The format does not appear to change because the condition is not met ie C2 is equal to B2.
(g) Now apply this format to all the cells in the cash flow by selecting C2, clicking on the **format painter** (standard toolbar) and then highlighting all of the cash flows from C2 to I7. (If you include B2:B7 they will all appear **Bold**. because they are all different from the respective cells in column A.)

3.5 Notation

Providing thorough explanations to accompany your worksheets is probably as important as the worksheet themselves, not just to explain the figures to those who may not be as familiar with appraisals and worksheets as yourself, but also to ensure that your message is not misunderstood if by unhappy chance there is an error. Good notes are the best defence against negligence.

To add a comment to one particular cell, right mouse click on that cell. From the pop up menu select **Insert Comment**. A floating text box will appear with placeholders for re-sizing the comment. The cursor is positioned so that you can start typing immediately.

	B	C	D	E	F	G	H	I
1	24 Mar 01	24 Mar 02	24 Mar 03	24 Mar 04	24 Mar 05	24 Mar 06	24 Mar 07	24 Mar 08
2	100,000	100,000	100,000	**120,000**	120,000	120,000	120,000	120,000
3	80,000	**110,000**	110,000	110,000	110,000	110,000	**125,000**	125,000
4	150,000	150,000	150,000	150,000	150,000	**175,000**	175,000	175,000
5	140,000	140,000	140,000	140,000	140,000	**165,000**	165,000	165,000
6	95,000	95,000	**120,000**	120,000	120,000	120,000	120,000	**145,000**
7	50,000	50,000	50,000	**70,000**	70,000	70,000	70,000	70,000
8					This is **Bold** because it			
9					increased on 24 March 2002			
10								
11								

If your copy of Excel has been saved with your name or your company name this will initially appear as the first line in the comment. Cells with comments have a small red triangle in the top right-hand corner.

3.6. Long formulae

3.6.1 *Splitting formulae*

There are many impressive worksheets in existence. Some are impressive principally for their long formulae. Although Excel is capable of handling extremely lengthy and complicated calculations in a single cell it is not always best practice to do this in your worksheet. Excel has an advantage over a calculator in providing many cells rather than a single display. Consequently it is possible in Excel to split a formula into smaller calculations and view the interim results. When complete we can hide the interim calculations from view.

Lengthy formulae often appear in appraisal worksheets, particularly cash flows, because they need to handle a choice of both when and which financial calculation to apply as well as the financial calculation itself. The formula for income and/or expenditure in a multi-tenanted cash flow is the most obvious and ubiquitous example. The rental income from one tenant will depend on several things eg rent reviews, voids, and whether or not a tenant uses a break clause. Calculating and entering all the individual income receipts independently can be a tedious process, and one which can conveniently be left to the computer if we create a formula to calculate :

(a)   At which point in time any of these events might occur.
(b)   What calculation to apply for each particular event.

A further complication may arise in that the calculation to perform on any particular event may itself change over time. A calculation that can cope with all of this will be intricate enough. Where there are several tenants the rental value, review pattern and predicted void length of each is likely to be different, adding yet more provisos to an already over-burdened formula.

The answer, as previously mentioned, is to break the formula down into more manageable parts[1]. The first step is to identify the component parts of the formula. The elements of any cash flow formula include both time and money. We illustrate a simple 10-year cash flow as our first example.

3.6.2 *Example*

The following cash flow represents a 10-year holding period for a property investment with a single tenant that produces an initial rent of £100,000, paid annually in advance, at an initial yield of 8%. Rent is assumed to grow at 3% pa, reviews are contracted to occur every 5 years.

Cell C7 refers to the rent in the inputs box ie = B3. Cell C8 contains a formula to calculate whether a rent review is due in 5 years from now. There are two quantitative elements of this formula – time and compound growth.

---

[1]   An alternative is programming in Visual Basic, and some suggestions to introduce you to the subject are made in Chapter 8. However programming models is not recommended as an alternative to the many bespoke programs available unless you have adequate resources.

	A	B	C	D	E	F
1	Rent	100000				
2	Yield	8.00%				
3	Growth	3.00%				
4	Review	5 yearly				
5						
6	Year	Purchase Price	Rent	Sale Price	Net cash flow	ERV
7	0	-1,250,000	100,000		-1,150,000	100,000
8	1		100,000		100,000	103,000
9	2		100,000		100,000	106,090
10	3		100,000		100,000	109,273
11	4		100,000		100,000	112,551
12	5		115,927		115,927	115,927
13	6		115,927		115,927	119,405
14	7		115,927		115,927	122,987
15	8		115,927		115,927	126,677
16	9		115,927		115,927	130,477
17	10			1,680,000	1,680,000	134,392
18						

Instead of trying to incorporate all of the calculations of both of these elements into one formula a running ERV has been calculated to the right of the cash flow. This is not a particularly important set of data of itself, and can be hidden from view[1], but having it around will make our rent formula simpler.

The following formulae are used :

(c) F7 refers to the original rent from the input box ie = B1.
(d) F8, contains the formula = F7 * (1 + $B$5). This is copied through to cell F17.
(e) C7:C16 combines these formulae :

The 'actual' rent column uses a formula that asks the question 'Is the rent period the review period?. If it is, use the reviewed rent in Column F, otherwise use the rent of the previous year. For this we use the IF function which has three arguments and the following syntax:

=IF(TEST, TRUE, FALSE)

The test will be a quantitative expression of the question 'Is the period the review period?' in this case is the relative cell in column A equal to 5? If the answer to the question is 'Yes' ie TRUE, copy the value in the relative cell in the ERV column. If the answer to the question is 'No', ie FALSE, copy the rent received in the previous year. Therefore in cell C12 we have the following formula:

=IF(A12=5,F12,C11)

---

[1] To hide this column, select the entire column by clicking on the column header F. Then select **Format Menu⇒Column⇒Hide**. Any references to this column will still work. To un-hide select the entire columns either side of the hidden column and then select **Format Menu⇒Column⇒Unhide**.

The result in this case is that in Year 5 the reviewed rent will be used, but in other years the rent will be unchanged.

A single formula in C8:C16 to carry out this operation would have been

$$=IF(A15=5,\$B\$3*(1+\$B\$5)^\wedge A11,C10)$$

While this formula is not particularly difficult, it would nevertheless be easier for another Excel user to understand the separate parts, particularly someone with comparatively little experience of Excel. If you use the Auditing facility it is much easier to see how the various parts have been combined.

3.7 A note on writing a single formula

If it is necessary to write the whole formula in a single cell you are less likely to make an error if you complete it a section at a time, working from the 'inside' out. To demonstrate this principle we take the formula for Years' Purchase Dual Rate which we discussed in *Excel for Surveyors*. This is conventionally written as:

$$\frac{1}{i + \dfrac{s}{(1 + s)^n - 1} * \dfrac{1}{1-t}}$$

Start with the expression $(1 + s)^n$ and check that your answer is correct according to either a valuation table or a pocket calculator. Continue by subtracting 1, enclosing in brackets and dividing it into s which should give the annual sinking field, which again you can check. Then multiply by $1/(1 - t)$; add s; enclose the whole in another set of brackets, and divide into 1. We emphasise the importance of checking your result at each step with a calculator or reference to tables, and perhaps reference to a colleague.

One of the main sources of error problems in constructing formulae is not getting brackets in the correct place. By compiling a formula in steps as described and checking result as you go along, the possibility of such errors is considerably reduced.

Chapter 4

Databases

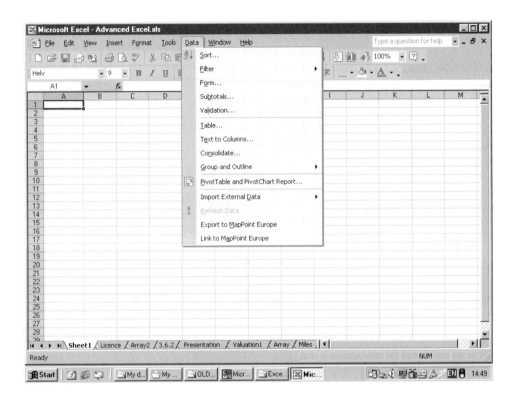

4.1 Databases – systems for cataloguing and finding information

A database is a structured method of storing data so that it can be retrieved easily. Some databases are created simply to provide single items of information as and when required – for example the standard telephone directory. Others, such as a card index, will allow the user to collate information in other ways – colour coding for example – so that a particular class of information can be readily identified.

We gave examples of how a simple database could be created in *Excel for Surveyors*. This involved records of sales of houses, including price, number of rooms, and other information. We also demonstrated how data could be sorted and filtered by various criteria.

We now consider a more complex example of a portfolio of many commercial properties in which individual valuations are made for the purposes of the annual report. The data is then assembled into a summary so that it can be used for detailed analysis in ways which we shall discuss later.

4.2 Designing the database

Before starting to compile our database the first obvious question to ask is, 'What is it for?' Purposes may vary from a simple facility to look up names and addresses to arranging detailed property data so that a portfolio analysis can be carried out. Before starting it is good practice to write down the name of each item of relevant data. This enables us to revise the eventual structure before getting too far into the compilation of the worksheet.

Having prepared our list of data items, and assuming that we are not importing them from an existing file, we enter each item name into the first row of a new worksheet. This will become the 'header' or name of each field for all subsequent operations.

If we later change our mind on the order in which the columns should appear all is not lost. One of the easiest ways to re-order columns is to insert a temporary row at the top of the worksheet and number the columns in the order into which we wish to change them, eg 1. 2. 5. 4. 3. Then go to the Sort Options dialogue box, select the whole of the data, and click the SORT button to sort on the temporary row, left to right, to change the order of the columns. Finally delete the temporary row. (Before you perform this operation, make sure that there is no other data below the main database, otherwise this will also be sorted.)

If one column will always contain a specific type of data it may help to avoid errors if you use **Format Menu⇒Conditional Formatting** ... to specify any conditions on data which may be entered. This also enables you to specify a particular format, font or cell colour if the data matches particular conditions, for example if it is a rent review due within the next six months.

If the data is to be used for analysis purposes consider whether you will be able to complete every cell for each record. Some analysis procedures can allow for missing data, but there is little point in providing for data if most entries are likely to be blank.

4.3 Flat files and relational databases

An important consideration which must be born in mind when considering Excel as a database is that an entire workbook must be loaded into memory before any operation can be performed. Even with the maximum of 65536 rows and 256 columns and sufficient memory to support all these cells there is obviously a limit on the amount of data which can be recorded. This is in contrast to a purpose-designed database program such as Microsoft Access™ or FileMaker in which the file is held on disk instead of being loaded into memory, and editing of data done directly to the file. (Consequences of this are that such a file is never expressly 'saved' and changes often cannot be undone.)

Another difference between Excel and a full database program is that while an Excel file can be used by more than one person at a time there are still some restrictions on the operations which can be performed.

Databases such as those mentioned are 'relational' – that is a complete database may consist of several files, each of which holds particular facets of information. A file which is in use by one operator can call other files such as client list, tenant list, or property list as required, and access can be restricted on a 'need to know' basis. Changes in the data on one file will be reflected immediately through the system. Excel does not have this facility though the 'Lookup' function can be used to transfer data from another file. However, the PC version of Microsoft Office includes Access which is a relational database, and which can easily be linked to provide calculation facilities to a full database system. The Macintosh version of Microsoft Excel will read FileMaker databases and conversely FileMaker will read Excel files, subject to the restrictions noted above.

Despite these disadvantages, an Excel database has many useful applications particularly where calculations are required. Access does not include any substantial calculation facilities, and those in FileMaker are less comprehensive than Excel. A database in Excel can be used directly for analysis, as we shall describe later.

There would normally be little point in importing data into Excel from other databases unless it is to be used for analysis which cannot be done in those other applications.

4.4 Sorting

4.4.1 Sorting rules

We considered the criteria for sorting in *Excel for Surveyors*, and pointed out that numbers could be sorted in ascending or descending order. Similarly, text can be sorted according to ascending (A–Z) or descending (Z–A) options. However problems may arise in the case of mixed alpha and numeric characters.

To understand the principles involved we must refer to the ASCII code (American Standard Code for Information Interchange). Each character is stored in a computer in a 'byte' which consists of eight bits (binary digits), each of which can have a value of 0 or 1. There are 256 such combinations for each byte and these represent the following characters (in this case using the Helvetica font on the Macintosh). The binary numbers are read left to right across each column and are the basis for sorting[1].

You will see that numerals appear on line 4 of the 16×16 matrix, followed by several other keyboard characters. Lines 5 and 6 contain capital letters and lines 7 and 8 lower case letters, together with some other keyboard characters. Lines 1–2 and 9–16 are disregarded for the purpose of sorting. So far as ASCII is concerned, the ascending order reads from left to right on each row, starting with row 1, column 1.[2]

Data including alphabetic and numeric characters will be stored as text, as will numeric data prefixed by a single quote ('). Data consisting only of numeric characters (including the decimal point and scientific notation if appropriate) will be stored as numbers.

---

[1] Machine code is also stored in byte form, and if you happen on such code you will find that it is completely unintelligible. Interfere with it at your peril!

[2] The decimal equivalents of these numbers are 0 to 255, and for example capital 'A' is decimal 65 or binary 010000001.

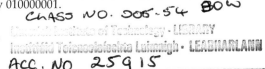

NUL	SOH	STX	ETX	EOT	ENQ	ACK	BEL	BS	HT	LF	VT	FF	CR	SO	SI
DLE	DC1	DC2	DC3	DC4	NAK	SYN	ETB	CAN	EM	SUB	ESC	FS	GS	RS	US
SPC	!	"	#	$	%	&	'	(	)	*	+	,	-	.	/
0	1	2	3	4	5	6	7	8	9	:	;	<	=	>	?
@	A	B	C	D	E	F	G	H	I	J	K	L	M	N	O
P	Q	R	S	T	U	V	W	X	Y	Z	[	\	]	^	_
`	a	b	c	d	e	f	g	h	i	j	k	l	m	n	o
p	q	r	s	t	u	v	w	x	y	z	{	|	}	~	DEL
Ä	Å	Ç	É	Ñ	Ö	Ü	á	à	â	ä	ã	å	ç	é	è
ê	ë	í	ì	î	ï	ñ	ó	ò	ô	ö	õ	ú	ù	û	ü
†	°	¢	£	§	•	¶	ß	®	©	™	´	¨	≠	Æ	Ø
∞	±	≤	≥	¥	µ	∂	∑	∏	π	∫	ª	º	Ω	æ	ø
¿	¡	¬	√	f	≈	∆	«	»	…	SPC	À	Ã	Õ	Œ	œ
–	—	"	"	'	'	÷	◊	ÿ	Ÿ	/	€	‹	›	fi	fl
‡	·	,	„	‰	Â	Ê	Á	Ë	È	Í	Î	Ï	Ì	Ó	Ô
	Ò	Ú	Û	Ù	ı	ˆ	˜	¯	˘	˙	˚	¸	˝	˛	ˇ

As a result of this, the sorting rules are as follows, assuming we choose ascending order.[1]

(a) Numbers are sorted to the top of the list in ascending order.

(b) Text entries are sorted in ascending order according to the ASCII table above. Thus entries starting with a space (line 3 column 1 of the table) will become the first of the sorted list, followed in succession by the characters on that line, then numerals (line 4), alpha characters, etc. Text preceded by a space will sort to the top of the list.

If your database includes a reference consisting of alpha characters followed by numerals it may be helpful to separate these into two columns so that the first sort can be by alpha and sub-sorted by numbers. We showed an example of this in *Excel for Surveyors* when we sorted house addresses by street name and then by house number.

(c) By default, Excel does not distinguish between upper and lower case alpha characters for sorting purposes. However this can be changed by clicking on the **Options** button in the **Sort** dialogue box and then clicking the **Case Sensitive** box.

---

[1] These rules for sorting have other uses outside Excel. For example, if the name of a file starts with a space, sorting files by name will place it at the top of the list. This can be very handy for finding a file which you refer to frequently.

(d) The default for sorting is by columns. If you need to sort by rows click the appropriate button in the Sort Options dialogue box.

4.4.2 List organization and formatting

The following points are suggested when you construct the list :

(a) Include only one list on each worksheet. Otherwise you may hide other data if you filter unless you position them below the list, or create other problems.

(b) It is recommended that you create a list sheet[1] if you expect that your list will contain more than 50 records (rows).

(c) Position other data above or below the list Avoid placing critical data to the left or right of the list as it may be hidden if you filter the list. It is safer not to enter any other data on the worksheet if you can avoid this – use another worksheet.

(d) Show all rows and columns. Ensure that any hidden rows or columns are unhidden before making changes to a list. When rows and columns are hidden data can be deleted inadvertently.

(e) Use formatted column headings. If you create a list with existing data, you can use the data's existing headings as column names or create new ones. Format the column names after you have created the list (bold, coloured, etc.).

(f) Format the entire list (including the column names) by clicking **Format Menu** ⇒**Autoformat** ... Alternatively you can format individual column names manually by selecting one and then clicking **Format Menu**⇒**Cells**. If you format manually, the format changes you make may not appear until you click outside the list frame.

(g) Do not type leading or trailing spaces before or after your data entries as these affect sorting and searching. If you need to indent text within the cell use **Format Menu**⇒**Cells**, click the **Alignment** tab and select a number in the **Indent** box. Alternatively use the Indent buttons on the toolbar.

4.5 Data validation and conditional formatting

You can apply data validation and conditional formatting either while creating a list or afterwards using **Data Menu**⇒**Validation**. Validation helps to ensure that the data entered into your list meets your criteria .

Data validation permits you to specify conditions about numbers which are acceptable. Validation allows other conditions about acceptable data, including text length, and also allows you to specify input messages and error messages.

You can specify other conditions by entering a formula as the formatting criteria. Click **Custom** ... in the first box, and then enter the formula in the box on the right. The formula must evaluate to a logical value of TRUE or FALSE.

---

[1] A list sheet is one which is dedicated to a single list.

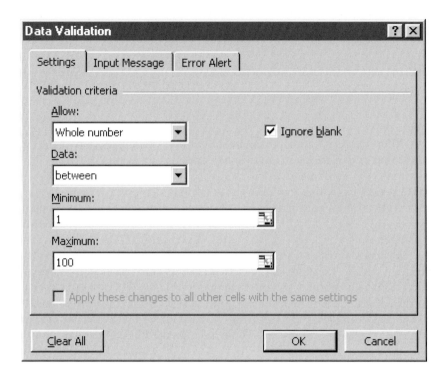

4.6 Using your database

Once your data has been entered (which may take quite a long time for a large database) it can of course be easily updated. Obviously a database which is out of date is probably worse than useless. We discussed the use of Forms for data entry and searching in *Excel for Surveyors*, and you may find that data entry is facilitated by using a form, particularly if the work is to be delegated to others.

Possible uses for your new database are legion, but we suggest the following:

(i) Looking up addresses, telephone numbers, and any other information about individuals (remembering of course the provisions of Data Protection legislation).
(ii) Selecting subsets of data by specified criteria.
(iii) Using in conjunction with Microsoft Word to mail merge documents
(iv) Statistical analysis.

4.7 Summarising data – a practical example

We take as our example a workbook which contains working details of a property portfolio and which we shall use again when we discuss the sharing of workbooks and portfolios in later chapters.

Each valuation is made on a single worksheet using whatever method of valuation is deemed appropriate. The final valuation and rental income for each property are set out in a defined format in the shaded cells at the head of each worksheet, which may or may not be in the same workbook as the eventual summary sheet.

	A	B	C	D	E	F	G	H
1				Reading			Berkshire	
2				Ref 1				
3		1999	2000	2001	2002	2003	2004	2005
4	Valuations	750,000	950,000	1,100,000	1,350,000	1,450,000		
5	Income	100,000	100,000	100,000	100,000	100,000		
6								
7	Region	Berkshire						
8	Tenant class	PLC						

Whatever the form of the actual valuation, for the purposes of this example, cells A1:I8 and A7:B9 (cells within the boxes) contain the same valuation details for each property, so that references in the final worksheet which forms the database of the portfolio are consistent.[1] Note that for the purposes of this example the valuation in F21 would be manually copied to G4 so that next year the same valuation format can simply be updated with revised inputs. (Extending to I8 allows for entries in future years.)

The heading of each valuation follows the above format, though the actual details of any valuation and its method will vary according to the property concerned. Providing that we have sufficient memory to load a workbook of 48 sheets plus those which we shall need for analysis, we can contain all these in one file. We take the above example as the first property to be included in the summary sheet, and details will be linked directly to the summary worksheet which appears below. For instance, the entry in Summary Cell D2 (this year's valuation) would read :

$$=Valuation\_1!F4$$

and similarly for the other links.

Below is the summary sheet showing details of all the valuations. (Except for the first property, all these are arbitrary valuations.)

Now that we have consolidated all relevant information into a single worksheet we can produce summary information. In the worksheet above we already have the totals of valuations for the past three years, but a more detailed analysis can include subtotals for each county, class of property and category of tenant. We shall consider this later when we discuss Maps.

If it is necessary for the valuations to be carried out by different valuers it is possible for each to make additions and alterations to their respective contributions to the workbook by making it **Shared**. Sharing workbooks will be considered later.

---

[1] In practice you would probably want to insert the full address in D1, but the town only is included here for the purposes of the example.

A	B	C	M	O	P	Q
0	Property	Location	Value 31 Mar 03	Rent 31 Mar 03	Use	Tenant Class
1	Reading	Berkshire	1,450,000	100,000	Office	PLC
2	Dover	Kent	5,161,000	300,000	Retail	Major PLC
3	London	Greater London	1,800,000	120,000	Office	Major PLC
4	Braintree	Essex	4,039,000	423,000	Office	Major PLC
5	Folkestone	Kent	174,000	18,000	Office	Major PLC
6	Margate	Kent	3,159,000	230,000	Retail	Government Dept
7	Watford	Hertfordshire	768,000	82,000	Office	Family Business
8	London	Greater London	862,000	91,000	Office	Family Business
9	St Albans	Hertfordshire	123,000	12,000	Office	Family Business
10	London	Greater London	3,547,000	431,000	Office	Government Dept
11	Southend	Essex	1,356,000	100,000	Industrial	Major PLC
12	Tring	Hertfordshire	1,537,000	147,000	Office	Major PLC
13	Reigate	Surrey	1,110,000	75,000	Industrial	Family Business
14	Letchworth	Hertfordshire	1,031,000	96,000	Industrial	Major PLC
15	London	Greater London	323,000	25,000	Office	Family Business
16	Maidenhead	Berkshire	3,603,000	275,000	Industrial	Major PLC
17	Reading	Berkshire	1,520,000	167,000	Industrial	PLC
18	London	Greater London	761,000	50,000	Industrial	PLC
19	Harlow	Essex	3,292,000	300,000	Industrial	Major PLC
20	London	Greater London	326,000	42,000	Retail	Family Business
21	Brentwood	Essex	312,000	43,000	Office	Family Business
22	Colchester	Essex	240,000	20,000	Office	Family Business
23	Wokingham	Berkshire	342,000	25,000	Industrial	PLC
24	London	Greater London	1,038,000	90,000	Industrial	Major PLC
25	Guildford	Surrey	1,270,000	162,000	Office	Major PLC
26	Newbury	Berkshire	515,000	62,000	Leisure	PLC
27	Camberley	Surrey	645,000	70,000	Retail	PLC
28	Canterbury	Kent	1,273,000	160,000	Retail	Major PLC
29	Woking	Surrey	186,000	21,000	Retail	Family Business
30	Ashford	Kent	2,835,000	150,000	Office	Major PLC
31	Welwyn G C	Hertfordshire	1,082,000	90,000	Leisure	PLC
32	Slough	Berkshire	2,089,000	125,000	Leisure	Major PLC
33	Hemel	Hertfordshire	1,672,000	245,000	Retail	Major PLC
34	Ramsgate	Kent	1,658,000	163,000	Retail	Major PLC
35	Chelmsford	Essex	438,000	59,000	Retail	Family Business
36	London	Greater London	664,000	89,000	Industrial	Major PLC
37	Basildon	Essex	787,000	50,000	Office	PLC
38	Leatherhead	Surrey	410,000	42,000	Industrial	Family Business
39	London	Greater London	1,667,000	100,000	Industrial	Major PLC
40	Chelmsford	Essex	58,000	1,000	Office	Family Business
41	Dorking	Surrey	820,000	97,000	Industrial	PLC
42	London	Greater London	746,000	73,000	Retail	Major PLC
43	Deal	Kent	140,000	10,000	Office	Family Business
44	Windsor	Berkshire	492,000	63,000	Retail	Family Business
45	Bracknell	Berkshire	1,200,000	90,000	Industrial	PLC
46	London	Greater London	4,060,000	380,000	Retail	Major PLC
47	London	Greater London	293,000	27,000	Office	Family Business
48	Stevenage	Greater London	1,521,000	100,000	Office	Major PLC

4.8 Arrays

An array is a particular form of database which enables us to perform several calculations and produce a single result. It is another way in which a computer can help us to do repetitive work.

As a very simple example of the use of an array, we assume a typical travelling claim in which the mileage for each day's work has to be multiplied by the cost per mile and totalled. While it is not difficult to do this by individually calculating the amount for each day, the entire operation can be carried out with a single formula.

	A	B	C	D	E
1	Date	Miles	Cost/Mile		
2	1 Mar	51	0.45		
3	2 Mar	74	0.45		
4	3 Mar	38	0.45		
5	4 Mar	95	0.45		
6	5 Mar	77	0.45		
7	6 Mar	84	0.45		
8	7 Mar	81	0.45		
9	8 Mar	72	0.45		
10	9 Mar	65	0.45		
11	10 Mar	46	0.45		
12				307.35	
13					

Click the cell in which you want to enter the array formula, D12

Type the array formula below, but do not include the braces. This obviously includes both the instruction to multiply each pair of values and sum the results.

$$=\{SUM(B2:B11*C2:C11)\}$$

Press **Crtl**, **Shift** and **Enter**. This will add the braces to indicate that it is an array formula, and carry out the necessary calculations.

The result appears as 307.35. This is a simple example and the same result could alternatively be obtained by using the SUMPRODUCT function ie = SUMPRODUCT (B2:B11, C2:C11).

(In previous versions of Excel arrays were used for conditional summation. These have been superseded by **Sumif** and other functions which we discuss elsewhere.)

Chapter 5

Sharing workbooks

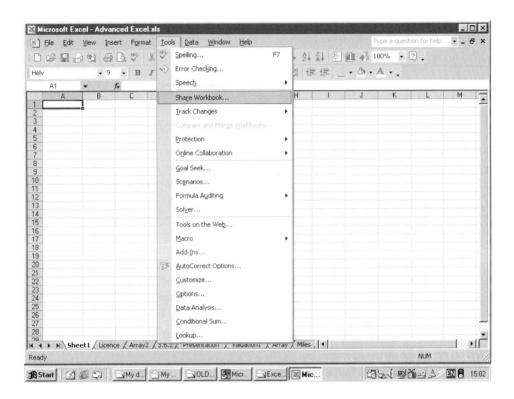

5.1 Working on a workbook simultaneously

Excel and other Office programs make it easy for you to share files with other people and work on the data together. If a workbook is shared it easy to review other people's files and allow others to review your files. We can share an Excel workbook with others in several ways, depending on requirements.

5.1.1 Copies for information

You can send copies of a workbook by attaching a copy to an e-mail, distributing your workbook by 'snail mail' on floppy or zip disks, storing your workbook on a network where others can download it, or publishing it on the Web if you want everyone to see it.

5.1.2 Review copies

If you want to invite review comments from several individuals, you can send separate copies of the workbook to each reviewer. If you have designated the workbook as **Shared**, as we shall shortly describe, you can merge any changes they made to those copies into a single workbook. You can accept and reject individual changes and review the history of all changes to the workbook. When you receive the marked-up copies of the workbook, you can review and incorporate the changes in each copy, or you can merge all the changes into one copy of the workbook.

5.1.3 Simultaneous users

If several users need to work in the same workbook simultaneously, we can save it as **Shared** and store it on a network[1]. Each user who opens the workbook can see the changes made by others. Provided that you have given the appropriate permissions, users can enter data, add and change formulas, and change formatting[2]. We shall consider one possible use of this facility later when we show how several surveyors can contribute towards the valuation of the portfolio of properties which we discussed in the previous chapter under the control of a supervising surveyor.

Excel identifies each user by a username. When users make changes to a shared workbook, Excel uses these usernames to identify their work. You or the individual user must set a username, by going to **Tools Menu⇒Options** and clicking **General**. Type the user name in the **User Name** box. Excel saves each user's personal filter and print settings for a shared workbook.

If several users make changes or add comments to the same cell in a shared workbook the text of all saved alterations and comments for the cell appears when each user saves the shared workbook. By default, you see changes made by other users who have saved their copies of the workbook whenever you save the shared workbook yourself. If you prefer, it can be set to save automatically at specified intervals. Each user can specify a different update interval.

You can view a complete list of changes to the workbook, including the names of the users who made the changes, data that was deleted or replaced, and information about conflicting changes.

5.1.4 Several users – one computer

Alternatively, you can set your own computer so that others can have access to particular files on a shared basis.[3] The principal user of the file can then declare it

---

[1]   You cannot save shared workbooks to Exchange public folders.
[2]   File sharing is not available in versions of Excel before 97/98. Some Excel commands and operations are not available in a shared workbook.
[3]   It is also possible for several individuals to share a single computer, each having their own files, user names and passwords. The computer will treat them as entirely independent users.

as shared and other individuals will have exactly the same privileges as if it were shared over a network.

5.2 Setting up a shared workbook

To set up a shared workbook we open it, click on **Tools Menu⇒Sharing**, and then click the **Editing** tab. Select **Allow changes by more than one user at the same time** check box, and then click OK and save. The word **[Shared]** will be attached to the filename. Copy or move it to a network location where other users can gain access to it.

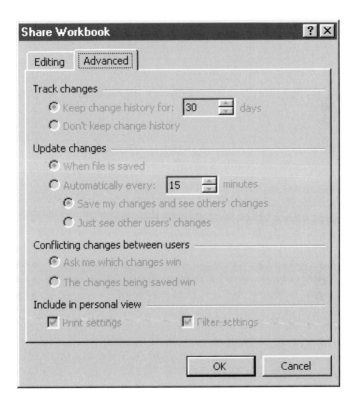

A maximum of 256 people can access a single copy of the workbook simultaneously. Clearly there needs to be some control over the changes, and therefore we can use **Change Tracking** to record who made which changes to the workbook and when, and **Change History** to manage the size of the file.

5.2.1 Setting up the CHANGE HISTORY

(a) Click the **Advanced** tab.
(b) Under **Track Changes**, click **Keep Change History for**, and then enter the number of days you want to maintain the change history in the **Days** box.

Excel maintains the **Change History** for only the number of days you specify – the default is 30 days.

(c)   Click OK, and save the file.

When you keep the change history for a workbook, Excel also turns on workbook sharing. Remember that some Excel commands and operations are not available in a shared workbook.

5.2.2 Giving others access to your workbooks

Turn on File Sharing and save, move, or copy the workbook you want to share to an empty folder or one that contains other workbooks you want to share. Click the folder that contains the workbook you want to share. Select the **Share this Item and its Contents** check box, and any other options you want.

By default, all users who have access to the network location where a shared workbook is stored have the same access to the shared workbook. If you want to prevent certain types of access to a shared workbook, you can protect it and the **Change History**.

5.2.3 Show or hide change highlighting and details

When you mark changes in a workbook by using the **Highlight Changes** command, you also turn on workbook sharing (if the workbook is not currently shared) and the **Change History**.

On **Tools Menu**⇒**Track Changes**, click **Highlight Changes**.

If you want to see the history of changes made by others (recommended) use **Tools Menu**⇒**Track Changes**⇒**Highlight Changes** and click the appropriate boxes. A blue triangle, similar to a Comments indicator, will appear in each cell which has been changed, and an explanation can be seen by positioning the cursor over the cell.

You can make it easy to see changes you and others make to a shared workbook by highlighting changes. Excel assigns a different colour to each user's changes. For example, a new row inserted by one user may be marked with green, and two cells changed by another user may be marked with yellow. You can also review the highlighted changes one at a time and then decide whether to accept or reject each one.

	A	B	C	D	E	F	G	H
1	Valuation							
2	Income		1,000					
3	YP 7. years	10%	4.8684	4,868				
4				**Philip Bowcock, 20/03/2003 15:08:**				
5	Reversion		5,000	Changed cell C5 from '6,000.00' to '5,000.00'.				
6	YP perp def 7. years	10%	5.1316					
7								
8								
9								
10								

If you do not want other users to turn off **Change History** or remove the workbook from shared use, type a password in the Password box, confirm it, and save the workbook. This shares the workbook and turns on the **Change History** but other users cannot then remove it from shared use or turn it off.

5.2.4 Instructions to users

It is often helpful to the other users to give instructions about the shared workbook by adding comments. This can be done in several ways :

(a) Select a convenient cell (such as A1), enter something like 'Read Me' in that cell, and attach a comment.

(b) Set aside a few lines at the top of a worksheet in which instructions can be given. Remember that there is a limit of 255 characters for any one cell.

(c) Create a sheet in the workbook specifically for instructions. To make it more readable you can turn off the gridlines and set all cells to **Wrap Text**. (Remember that the limit of 255 characters still applies.)

If you copy the shared workbook to a network location, make sure any links to other workbooks or documents are intact. Use the Links command on the Edit menu to make corrections to the link definitions.

5.3 The history of changes to a shared workbook

Change History enables you to view information about changes made, including the author of the change and the data that was entered and possibly replaced later. Ensure that the **Track Changes While Editing** check box is selected.

5.3.1 Viewing changes to a workbook

You can view the history in two ways: highlighted on the worksheet with details displayed when you rest the pointer over a changed cell, or listed on a separate History worksheet. When you view the history on a separate worksheet, you can filter the information in more than one way. For example, you can filter to find different types of changes, such as those made by different users on a particular date.

Display the History worksheet instead of highlighting changes when you want to view all the changes ever made to the worksheet, you want to view a large number of changes, you want to view information about how conflicting changes were resolved, or you want to print the change information.

5.3.2 Tracking changes

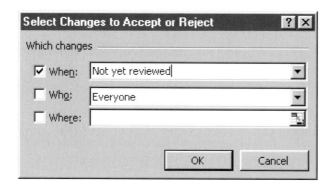

(a) To view changes made by a specific user, select the **Who** check box, and then click a user on the **Who** pop-up menu. Similarly, to view all changes to a specific range of cells, select the **Where** check box, and then enter a range reference in the **Where** box, or select a range on the worksheet.

(b) To view the details about changes on a separate History worksheet, select the **List Changes on a New Sheet** check box.

5.3.3 Limits on change highlighting

When you use the **Highlight Changes** command (**Tools Menu**⇒**Track Changes**), Excel highlights the following changes:

(a) Changes to cell contents, including moved and pasted contents
(b) Inserted and deleted rows and columns.

5.3.4 Keeping personal view and print settings for a shared workbook

Each user can set independent view and print options. To do this, go to **Tools Menu**⇒**Share Workbook**, and then click the **Advanced** tab. To save personal print settings, check that the **Print Settings** check box is selected under **Include in**

Personal View. Print settings include page breaks, print areas, settings you make in page break preview and settings you make in the **File Menu⇒Page Setup** dialog box.

To save any settings you make by using the commands under **Data Menu⇒ Filter**, including filtering done using the AutoFilter command, ensure that the **Filter Settings** check box is selected under **Include in Personal View**.

5.4 Things you cannot do when you share a workbook

Some features of Excel are not available in a shared workbook. If you need to use these features, do so before you share the workbook, or alternatively temporarily remove the workbook from shared use. They are listed in the Help file under 'Limitations of shared workbooks'.

5.5 Merging copies of a workbook

As indicated at the beginning of this Chapter, you may have sent copies of your workbook to other users for them to make amendments. On their return you can merge the amended copies. It is essential that each copy of the shared workbook maintains the change history from the day when you created it until the day of merging. If the number of days you originally specified has been exceeded you can no longer merge the copies. If you are unsure how long the review process will take, ensure you maintain the change history for a sufficient number of days, or enter a large number of days, such as 1,000.

5.5.1 Merging changes from multiple copies of the same workbook

Open the copy of the shared workbook (the 'editorial' workbook) into which you want to merge changes from another workbook file. Go to **Tools Menu⇒Merge Workbooks** and in the **Select File to Merge into Current Workbook** dialog box, click a copy of the shared workbook that contains the changes to be merged, and OK. Repeat with each copy of the shared workbook.

If a cell contains a comment, the comment will include the name of the person who inserted it. If a cell contains comments from more than one person, they will all appear in the comment box for that cell.

5.5.2 Updating a shared workbook

Each user of a shared workbook can set independent options for the frequency of receiving other users' changes. Open the shared workbook, click **Tools Menu ⇒Share Workbook**, and then click the **Advanced** tab. To see other users' changes each time you save the shared workbook, click **When File is Saved** under **Update Changes**. To view other users' changes periodically, click **Automatically Every** Under **Update Changes**, and the frequency in the **Minutes** box, and then click **Just see Other Users' Changes**.

To save the shared workbook each time you get an update so that other users can see your changes, click **Save my Changes and see Others' Changes**.

5.6 Conflicting changes

Obviously there needs to be some control over situations where conflicting changes are made to a workbook, and someone, presumably the original creator of the workbook, will need to act as 'editor' by retaining control and resolving conflicts. This is obviously a reason for limiting other users' facilities and using password access.

By default, if the changes you are saving conflict with changes made by another user, Excel displays the **Resolve Conflicts** dialog box so that you can decide whether to save your changes or keep the changes made by others. Alternatively you can choose to save your changes automatically instead of reviewing conflicting changes. If you decide to review the conflicting changes, you can view information about each change and decide which ones to keep or discard.

When you save a shared workbook you may see the **Resolve Conflicts** dialog box. For each change, read the information about your change and the conflicting changes made by others, and click either **Accept Mine**, or **Accept Other**. Alternatively you can keep all of your remaining changes or all of the other users' changes, by clicking **Accept all Mine** or **Accept all Others**.

If you do not want Excel to display the **Resolve Conflicts** dialog box when you save a shared workbook, you can set the option to save your changes automatically instead of reviewing conflicting changes. If you do not want to see the Resolve Conflicts dialog box when you save a shared workbook, click **The Changes being Saved Win**. The stored change history is erased the next time you save the shared workbook. However, if you turn on the **Change History** again, the last set of changes that were saved before you turned on the change history is included in the new change history.

5.7 Unsharing a workbook

If you decide that you no longer want others to make changes to a shared workbook, you can open and work in the workbook as its only user. However when you remove a workbook from shared use you disconnect all other users from the shared workbook, **Change History** is turned off, and the stored change history is erased so that you can no longer view the history or merge this copy with other copies of the shared workbook.

Caution: To ensure that others do not lose work in progress, make sure that all other users have been notified so that they can save and close the shared workbook before you remove it from shared use.

(a) Make sure that protection of the workbook is turned off by selecting **Tools Menu⇒Protection**.
(b) Go to **File Menu⇒Share Workbook**, and click the **Editing** tab.
(c) Ensure that you are the only person listed in the **Who has this Workbook Open Now** box. If other users are listed, they will lose any unsaved work.
(d) Clear the Allow changes by more than one user at the same time check box, and click OK.
(e) When prompted about the effects on other users, click YES.

To remove a user who appears to be connected to a shared workbook but is no longer working in the workbook or whose network connection is broken, go to **Tools Menu⇒Share Workbook**, and click the **Editing** tab. In the **Who has this Workbook Open Now** box, click the name of the user you want to disconnect, and click **Remove User**.

If a user no longer needs to work on your shared workbook, you can reduce the size of the file by deleting that user's personal view settings. Click **View Menu ⇒Custom Views**, click the user's view, and then click **Delete**. Of course, that user's unsaved work will be lost.

Chapter 6

Maps

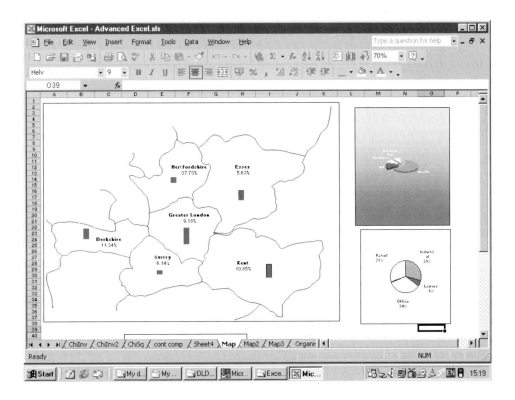

6.1 Creating maps in Excel

It not infrequently happens that a report to an investor client can be enhanced with a graphical representation of information, and some investors request this. There are numerous styles of presentation, including bar charts, trend lines and pie charts with dozens of variations of detail. We considered some of these in *Excel for Surveyors*. One method where data is distributed in geographical regions, is to add a chart to each region so that the overall perspective can be seen. In this chapter we will show how such a map can be produced in Excel.

6.2 Importing and exporting data

The data on which a map (or any other analysis) is to be based may be held on other databases, and it will therefore be necessary to transfer it into an Excel file. Before discussing the preparation of maps we will consider two examples of this procedure, though of course there are files held in many other database formats.

6.2.1 Importing data from an Access file into Excel for PC

To transfer an Access file into Excel it must first be exported from Access as a Text file. Open the Table of the Access file and go to **File Menu⇒Export**. Enter a file name and select a file name (in this example 'Excel version') and the Type as 'Microsoft Excel 97–2000'.

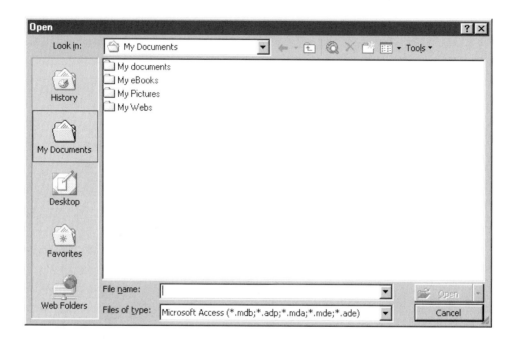

A new file will appear in the directory with an Excel icon and this can be opened in the normal way.

With Excel open, go to **Data Menu⇒Get External Data⇒Import Text File** and open the new file. If it has text file format (as may happen) it will be necessary to use **Data Menu⇒Text to Columns** to correctly separate the data into separate fields, as we discuss shortly.

If numerical data appears in parenthesis (' ') these must be removed, otherwise the data will be regarded as text. To do this, select all the columns and use **Edit Menu⇒Replace** (**Cmd** F and click the **Replace** tab) and **Replace All** and replace the quotes with a blank, ie, do not insert anything in the **Replace With** box. The data will now become standard numeric.

The items of data will normally be delimited by tabs, commas or spaces, so select the appropriate option and follow the instructions.

(Access can read an Excel file and link to it if it is necessary to reverse the transfer process. If there are several workbooks in the file you will be asked which one is to be linked.)

6.2.2 Importing data from FileMaker into Macintosh Excel

Macintosh users who are looking for a database might well consider FileMaker which has been designed to have close compatibility with Excel. It is a relational database which is easy to set up, and is also fully compatible with Windows.

Using System X (System 10), importing data from a FileMaker file into an Excel worksheet is incredibly easy. Simply drag the file onto the Excel icon and a new Excel file will be created. The Import Wizard dialogue box will ask which fields you want to import – follow the on-screen instructions to make your selection. Remember that there are restrictions of 65,536 rows and 256 columns in a worksheet, and there is always the limit of available memory.

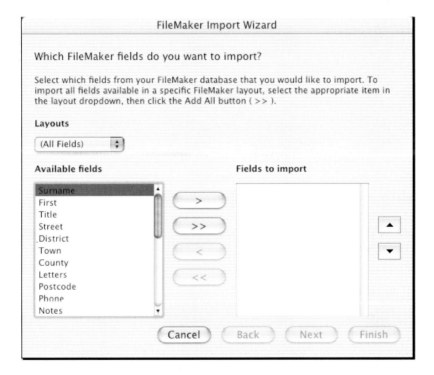

The 'Next' button allows you to specify criteria to select a subset of the whole file. For example you may wish to import only records of people in London. When you are done click 'Finish' and the data will be transferred. Use the side arrows if you need to change the field order in the new worksheet. Similarly you can click on the single or double reverse arrows to deselect one or all of the fields.

The reverse process of creating a FileMaker file from an Excel worksheet is similar – drag the workbook onto the FileMaker icon and follow the instructions. Only one worksheet can be transferred so the Import Wizard will ask for this information. Select your fields and click on the arrow.

You should appreciate that you have created a new file, and the original FileMaker file remains intact.

Which FileMaker records do you want to import?

Reduce the amount of records you import by specifying up to three criteria. Select a field and value for each criteria. To return all records, leave all criteria blank.

Criteria 1

| Pick a field... | ⬍ | = | ⬍ | |

Criteria 2

⦿ And   ◯ Or

| Pick a field... | ⬍ | = | ⬍ | |

Criteria 3

⦿ And   ◯ Or

| Pick a field... | ⬍ | = | ⬍ | |

(Cancel)   (Back)   (Next)   (Finish)

(Note that data will not be saved back to the original FileMaker file – if you want your alterations to be merged back into the original file you will need to do this within FileMaker.)

6.3 Text to columns

It may happen that when importing a file into Excel, text is put into one cell when you wish it to be divided into several. For example, suppose a column of names is imported in the style of 'Mr John Smith' and you want this to be divided into three columns of Title, First Name and Surname. Proceed as follows.

Caution: be sure that there are sufficient blank columns to the right of the data, otherwise other data may be overwritten.

Select all the names to be converted and go to **Data Menu**⇒**Text to Colums** ... The following dialogue appears, and you now have two options. In this case, as the words are separated by spaces we select **Delimited** and after clicking **Next** select **Space**. In other cases where the fields are of constant width we select **Fixed Width**, and on clicking **Next** Excel will propose fixed divisions. These can be adjusted manually if necessary.

In either case, the **Next** button takes you to further options where you can specify whether a column is to be formatted as a date or text, and in the case of numbers, how the data is to be translated if it includes decimal points and thousands separators.

Click on **Finish** and the original data will be divided into separate columns.

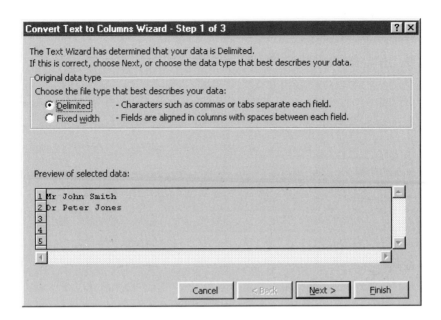

6.4 Preparing the data

We will use the data which we considered previously in our discussions of Sharing and Databases. Having created our portfolio of valuations of investment properties (which may have been done by several different valuers who have added their contribution to a shared file as we discussed in the last chapter), we are now instructed to prepare a map showing charts of the values in each of the counties concerned. The charts are to show capital values for the last three years and are to be shown on the respective counties on the map. We will therefore require subtotals for each of the counties concerned, but we will also take the opportunity to create subtotals for the different types of property and class of tenant.

We start by setting out the following table headings below the data on our summary worksheet showing the following:

52				
53	Surrey			
54	Essex			
55	Berkshire			
56	Kent			
57	Hertfordshire			
58	Greater London			
59				
60				
61	Industrial			
62	Leisure			
63	Office			
64	Retail			
65				
66				
67	Major PLC			
68	PLC			
69	Family Business			
70	Government			
71				

6.5 Calculating the subtotals (SUMIF)

For this purpose we use the **Sumif** function to obtain subtotals of the data which meet our specified criteria. Start by selecting Cell C53, and go to **Insert Menu** ⇒**Function**, and select the **Sumif** function in the usual way. For the **Range** box drag over the Location column (C2:C48); in the Criteria box enter the name of the county which we inserted in Column A (A58); and in the **Sum_Range** box drag over all the valuations of the 1990 ((F2:F48), as shown below. Click OK.

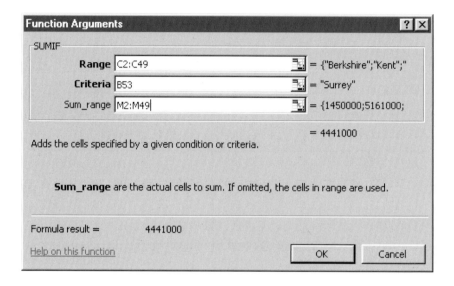

Repeat the operation for each county, and also for the type of property and class of tenant, and the results will be as below. We show some of the formulae which will have appeared in the A53:A58 as a result of using the **Sumif** function and then the full summary calculations. In each case we have also included a grand total to check that it matches the totals of D50:F50.

	A	B	K	L
1	0	Property	Value 31 Mar 01	Value 31 Mar 02
52				
53		Surrey	=SUMIF($C$2:$C$49,$B$53,$K$2:$K$49)	=SUMIF($C$2:$C$49,B53,$L$2:$L$49)
54		Essex	=SUMIF($C$2:$C$49,B54,$K$2:$K$49)	=SUMIF($C$2:$C$49,B54,$L$2:$L$49)
55		Berkshire	=SUMIF($C$2:$C$49,B55,$K$2:$K$49)	=SUMIF($C$2:$C$49,B55,$L$2:$L$49)
56		Kent	=SUMIF($C$2:$C$49,B56,$K$2:$K$49)	=SUMIF($C$2:$C$49,B56,$L$2:$L$49)
57		Hertfordshire	=SUMIF($C$2:$C$49,B57,$K$2:$K$49)	=SUMIF($C$2:$C$49,B57,$L$2:$L$49)
58		Greater London	=SUMIF($C$2:$C$49,B58,$K$2:$K$49)	=SUMIF($C$2:$C$49,B58,$L$2:$L$49)
59			=SUM(K53:K58)	=SUM(L53:L58)
60				
61		Industrial	=SUMIF($P$2:$P$49,B61,$K$2:$K$49)	=SUMIF($P$2:$P$49,B61,$L$2:$L$49)
62		Leisure	=SUMIF($P$2:$P$49,B62,$K$2:$K$49)	=SUMIF($P$2:$P$49,B62,$L$2:$L$49)
63		Office	=SUMIF($P$2:$P$49,B63,$K$2:$K$49)	=SUMIF($P$2:$P$49,B63,$L$2:$L$49)
64		Retail	=SUMIF($P$2:$P$49,B64,$K$2:$K$49)	=SUMIF($P$2:$P$49,B64,$L$2:$L$49)
65			=SUM(K61:K64)	=SUM(L61:L64)
66				
67		Major PLC	=SUMIF($Q$2:$Q$49,"Major PLC",$K$2:$K$49)	=SUMIF($Q$2:$Q$49,B67,$L$2:$L$49)
68		PLC	=SUMIF($Q$2:$Q$49,B68,$K$2:$K$49)	=SUMIF($Q$2:$Q$49,B68,$L$2:$L$49)
69		Family Business	=SUMIF($Q$2:$Q$49,B69,$K$2:$K$49)	=SUMIF($Q$2:$Q$49,B69,$L$2:$L$49)
70		Government	=SUMIF($Q$2:$Q$49,B70,$K$2:$K$49)	=SUMIF($Q$2:$Q$49,B70,$L$2:$L$49)
71			=SUM(K67:K70)	=SUM(L67:L70)

The following is the result:

Surrey	3,957,000	4,184,000	4,441,000	6.14%
Essex	9,438,000	9,961,000	10,522,000	5.63%
Berkshire	9,277,000	10,069,000	11,211,000	11.34%
Kent	12,364,000	12,990,000	14,400,000	10.85%
Hertfordshire	5,649,000	4,511,000	6,213,000	37.73%
Greater London	15,064,000	16,127,000	17,608,000	9.18%
	55,749,000	**57,842,000**	**64,395,000**	11.33%
Industrial	16,289,000	17,230,000	18,814,000	9.19%
Leisure	3,120,000	3,329,000	3,686,000	10.72%
Office	18,665,000	18,791,000	22,079,000	17.50%
Retail	17,675,000	18,492,000	19,816,000	7.16%
	55,749,000	**57,842,000**	**64,395,000**	11.33%
Major PLC	36,740,000	37,443,000	42,486,000	13.47%
PLC	7,806,000	8,421,000	9,122,000	8.32%
Family Business	5,255,000	5,623,000	6,081,000	8.15%
Government	5,948,000	6,355,000	6,706,000	5.52%
	55,749,000	**57,842,000**	**64,395,000**	11.33%

6.6 Microsoft MapPoint

This is a separate Microsoft application with which a wide variety of maps can be plotted from the data we have assembled above. The European version shows a great deal of information including, for this country, population and income bands. It is possible to create regions ('territories') and generate graphics from Excel data, but there appear to be some limitations, for example the names of towns and other information cannot be removed, which may not be what you want.

6.7 Custom maps

Many situations may require a map designed for the purpose, and we now demonstrate how a custom map can be created which is dynamically linked to data in another worksheet. The initial setting up does require some detailed work, but thereafter will be available for subsequent reports.

6.7.1 Preparing the worksheet

We shall need a new worksheet for the map, so proceed as follows :

(a) Create the new sheet with **Insert Menu⇒Worksheet**.
(b) Go to **Tools Menu⇒Options**, **View**, and clear the **Gridlines** box.

(c) Open the Drawing toolbar by clicking on the **Drawing** button or using **View Menu⇒Toolbars⇒Drawing**.

(d) Select the **Rectangle** button and drag across the worksheet to create a map area on the visible area of the worksheet – avoid scrolling if possible as can lead to problems of alignment. If your map is too large to fit on the screen you can use **View Menu⇒Full Screen** or use the Percentage button to reduce the size of your map.

(e) While the **Rectangle** is selected go to **Format Menu⇒Autoshape**, click the **Colours and Lines** tab, and under **Fill** select **No Fill**.

We now have a rectangle in which we can draw the map – in this case the boundaries and names of the counties. However any data entries in the worksheet cells will still be visible because we selected **No Fill**. More importantly, we can add charts to the drawing area. You should appreciate that the drawing is superimposed over the worksheet cells, and drawings can be superimposed over each other. Drawings can be moved relatively forwards or backwards on the worksheet depending on how you wish to view or modify them.

6.7.2 Creating the map

This is the tricky part of the operation. We use the mouse for tracing the county boundaries as follows[1].

(a) Scan a copy of the map area from an appropriate map and save it as a picture of size similar to that of the worksheet drawing area, and in a format which Excel can import. You can use any map which shows your area boundaries but it is easier if you first prepare an outline map showing only the actual boundaries which you want on your final production. Depending on the application, you may alternatively be able to scan and do a straight copy and paste. Otherwise save it as a picture and use **Insert Menu⇒Picture⇒From File** to copy the map into Excel. Adjust it to fit your defined map area, stretching or shrinking if necessary by dragging the bottom right corner handle.

    (*Tip*: if you use a light colour for your outline it is much easier to trace over it.)

(b) Open the Drawing toolbar either by clicking on the drawing button (if it is visible) or by **View Menu⇒Toolbars⇒Drawing**. Click on the Drawing button (the first one) and select **Arrange⇒Send to Back** so that the map rectangle on which you will trace the county outlines is in front of it.

(c) On the **Drawing** toolbar, click on **Lines** and select **Freeform** or **Curves** to show a drawing pencil, and trace round the outlines of the counties with this. You will almost certainly be drawing more than one line, and if you 'tear off'

---

[1] As an alternative to using the mouse, if you have a Wacom tablet you can place an actual map on the tablet and trace with the stylus, but you should be aware that unless you are reasonably expert in using the tablet it is difficult to guarantee consistency between horizontal and vertical scales.

the **Lines** box and drag it to convenient part of the screen it will save continually opening it from the toolbar.

(d) You trace by clicking on the outline at intervals with either your mouse or tablet pen and the points are automatically joined. The difference between **Freeform** and **Curves** is that the former joins points by straight lines and the latter calculates smooth curves.

(e) If there are errors in the lines of your drawing you can edit them. Click on the line to be altered (which will then show a rectangle around it) and go to the **Drawing Toolbar⇒Draw⇒Edit Points**. All the change points of the line will appear and can be dragged to new positions. To finish, click outside the rectangle. However for the purpose under discussion great precision is probably not necessary – the map is to show something about property values and not the quickest way to a channel port if you suddenly have an urgent need to leave the country.

When you are satisfied with your outline map select your original scanned picture, drag it away from your work, and delete it.

You will almost certainly by now have numerous different lines as it is very unlikely that you will be able to draw the entire boundaries of all areas in one operation. To save the individual lines from being accidentally moved we now group all of them into one unit. Select all lines including the border of the map by clicking on the selection arrow and dragging over the entire area, and then click on **Drawing Toolbar⇒Draw⇒Group** to create a single unit. Should you need to alter any line you can always use the **Ungroup** command, make the alteration, and **Re-Group**.

You can now revert to using the worksheet as normal. Clicking on any part which is not a line will select the normal cell and we can add county names or titles or any other information into the cell in the usual way, when they will show through the map.

6.7.3 *Attaching charts to the counties*

We now go to our summary data again to prepare our first chart, to which we have added one further column showing the percentage increase from 2000 to 2001. The purpose of this will appear later.

We can select any county to start with, but it will be most convenient to start with the one having the highest values, so we will choose Greater London and proceed in the normal way to create a chart:

(a) Click the **Chart** button and select **Column** and **Next**.
(b) Click on the Summary worksheet and select cells C58:E58 (or whatever your own cell references are) and **Next**.
(c) In Step 3 clear all boxes in all tabs as any text will be too small to read on the map and will just look untidy.
(d) Set the graph to be placed as Object in your Map worksheet using the side arrow, and **Finish**.
(e) The chart is now added to the Map worksheet. Double click on the Y-axis (vertical) scale to show the Format Axis dialogue box again.

51				
52				
53	Surrey	3,957,000	4,184,000	4,441,000	6.14%
54	Essex	9,438,000	9,961,000	10,522,000	5.63%
55	Berkshire	9,277,000	10,069,000	11,211,000	11.34%
56	Kent	12,364,000	12,990,000	14,400,000	10.85%
57	Hertfordshire	5,649,000	4,511,000	6,213,000	37.73%
58	Greater London	15,064,000	16,127,000	17,608,000	9.18%
59		**55,749,000**	**57,842,000**	**64,395,000**	11.33%
60					
61	Industrial	16,289,000	17,230,000	18,814,000	9.19%
62	Leisure	3,120,000	3,329,000	3,686,000	10.72%
63	Office	18,665,000	18,791,000	22,079,000	17.50%
64	Retail	17,675,000	18,492,000	19,816,000	7.16%
65		**55,749,000**	**57,842,000**	**64,395,000**	11.33%
66					
67	Major PLC	36,740,000	37,443,000	42,486,000	13.47%
68	PLC	7,806,000	8,421,000	9,122,000	8.32%
69	Family Business	5,255,000	5,623,000	6,081,000	8.15%
70	Government	5,948,000	6,355,000	6,706,000	5.52%
71		**55,749,000**	**57,842,000**	**64,395,000**	11.33%
72					

(f) Double click on the Y-axis, set the minimum to 0 and the maximum to something above the maximum – in this case 20000000. Clear the Auto boxes and OK.

(g) With the chart selected, click on **Chart Menu⇒Chart Options** ... Under the **Axes** tab now click the Y-axis box to delete it again and OK.

(h) We now have a chart showing how the total value of the portfolio has changed in Greater London over the last three years, but we still need to do some editing of the graph before it can be added to the map.

(i) Double click on one of the columns to bring up Format Data Series.

(j) Click on one of the columns to see the **Format Data Series** panel.

(k) Under Patterns Border click None

(l) Under **Options** click **Vary Colours by Point** and set **Gap Width** to zero, and OK. If you would prefer a different colour for one of the bars click on it, and if this selects all the bars click again to select that one only. Double click again and select the desired colour from the patterns shown. Set the Gap Width to zero and OK.

(m) Double click on the chart border and in the **Patterns** box click **None**.

(n) Click on the grey chart area to show the **Format Plot Area** panel and click **None** on the **Border** and **Area** radio buttons.

(o) Our bar chart now shows nothing but the coloured value bars. The final step is to reduce the size of this to a convenient size for adding to the map.

(p) Hold down the **Ctrl** key, click the chart area and select **Format Menu⇒ Format Object** to see the dialogue box.

(q) We now need to resize the chart to fit over the map, and it is important to remember to distinguish between the Chart Area – the outside box, and the Plot Area within it.

(r) With the Chart unselected, hold down the **Ctrl** key and then click on the

Chart. The handles this time will be small circles. Go to **Format Menu⇒ Format Object** and under **Size** set the height and width.

(s)   Keeping the Chart selected, open the Drawing toolbar and click on **Draw⇒ Order⇒Send to Back**. This will ensure that the lines of the county boundaries will not be hidden by the chart.

(t)   Finally, drag the chart to Greater London and place it in a convenient position.

(u)   Add similar charts to each of the other counties. However we can use our work to date to shorten the process and ensure that all charts are of the same scale. Proceed as follows :

(v)   Hold down the **Ctrl** key, click the Greater London chart and **Copy**.

(w) Paste the chart to another county, for example Surrey, keep it selected, and go to **Chart Menu⇒Source Data**.

(x)   Delete the **Data Range** and insert a new range by dragging over the data for Surrey and OK (the source worksheet will have appeared at this point).

The amended chart will now show the Surrey data. Note that by copying the London chart the relative sizes of the bars will be preserved. If we started again for each county in turn it would be more difficult to keep proportions identical.

The last action is to add county names (if we have not already done this) and the percentage increase information from the data worksheet. These are entered into convenient worksheet cells, linking the percentages to the data. We have also formatted the county names in bold and centred them in their cells. If necessary the charts can be moved to avoid overlap.

We now have our final map with all counties completed.

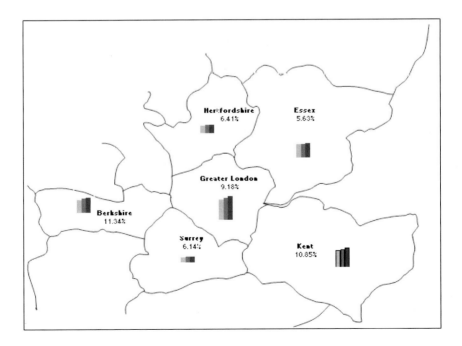

Now that the map has been created, adding data for subsequent years and updating the map charts is a very easy matter.

Of course there are so many other possibilities for using different forms of chart, or different information such as rents, population, or average value of properties that it would be impossible to discuss every alternative, so we leave this to your imagination and the clients' objectives.

Chapter 7

Protecting your workbook and setting passwords

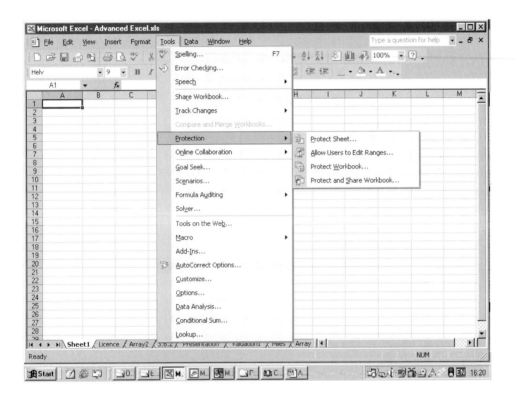

You may wish to restrict the access of other users to your workbook[1]. There are several possibilities. You can prevent access to the entire workbook, to individual worksheets, to particular cells or objects, or to formulae. You can also make the workbook or worksheets read-only, preventing changes by others.

Protection works in two stages. Individual cells and objects including charts and pictures are marked as locked by default. However locking has no effect unless the locked items are subsequently protected.

When you protect the worksheet any cells and objects which are not unlocked will be protected and cannot be changed.

---

[1]   If the workbook is shared these can only be applied before sharing is turned on, or by removing sharing temporarily.

7.1 Specifying password access to an entire file (a wise precaution)

This prevents unauthorised persons opening it. To set a password, use the **File Menu⇒Save As** command; then click on **Tools** and **General Options**. Enter your password to open, and if it is to be read-only, a password to modify. Click OK and you will be asked to enter the password again for safety. **Remember that when you assign a password to any operation, you should write it down and keep it in a secure place. If you lose the password, you lose the lot!**

7.2 Locking and unlocking individual cells or worksheets

You can lock or unlock single cells or groups of cells, individual objects (charts, pictures, drawings etc), specified rows or columns, whole worksheets or the entire workbook. The last named is additional to the password protection discussed above which prevents any unauthorised person from opening the workbook at all.

As mentioned above, cells and objects are locked by default, and therefore we need to specify any items which can still be changed when protection is turned on as we shall describe later. To do this, select the worksheet you want to protect and then the cells, formulas or graphic objects that you want to be able to change. Go to **Format Menu⇒Cells** ... and the **Protection** tab, or in the case of objects, the **Properties** tab, and clear the **Locked** box.

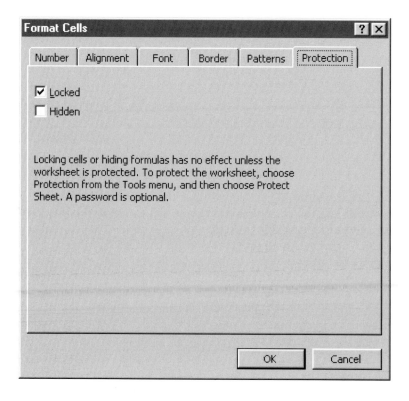

To stop other users from seeing particular rows or columns of a worksheet, select those rows or columns, go to **Format Menu⇒Row** or **Column**, and click **Hide**. If a cell containing a formula is locked and hidden the formula will not be visible but the result of the formula will still appear.

Another way to prevent others from seeing cells is to format the font in the same colour (by default this will be white) as the cell before protecting it, though the contents of the cell will still be visible in the formula bar if the cell is selected. This is however a useful technique if the worksheet is to be printed..

If you want to prevent others seeing the worksheet at all, hide it with **Format Menu⇒Sheet⇒Hide**.

7.3 Protecting locked elements

The second stage is to apply protection by going to **Tools Menu⇒Protection** and **Protect Sheet** or **Protect Workbook**, as appropriate.

To prevent changes to cells on worksheets or to data and other items in charts, and to prevent viewing of hidden rows, columns, and formulas, go to **Tools Menu ⇒Protection** and select the **Contents** check box. You then have the options to prevent changes to cells, to graphic objects or charts, and the definitions of scenarios on a worksheet by checking the appropriate boxes.

Your selection can be protected with a password.

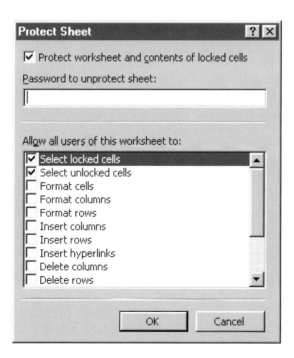

Now that your worksheet is protected the cells that you unlocked are the only cells that can be changed. You can move between unlocked cells on a protected worksheet by clicking an unlocked cell, and then using **Tab** to move to others.

Note that when you unhide a worksheet with the password that password is forgotten and to rehide it you must go through the password setting procedure again. Be sure that you use the same password, or write down the new one.

7.4 Consequences of protection

(a) If you hide a formula the result will still be shown, but not the underlying formula.
(b) When you select the Objects check box in the Protect Sheet dialog box you cannot do the following unless protection is turned off first or you unlocked the objects before you protected the worksheet:
    (i) Change graphic objects, including embedded charts, shapes, text boxes, and controls. For example, if a worksheet has a button that runs a macro, you can click the button to run the macro but you cannot delete the button. However, a protected chart will update if you change the source data.
    (ii) Enter or edit comments.
    (iii) Create or change scenarios, or view hidden scenarios. You can however edit the values in the changing cells, if the cells are not protected.

7.5 Limiting changes to the entire workbook

Go to **Tools Menu**⇒**Protection**, and **Protect Workbook**. Here there are two options:

(a) To protect the structure of a workbook so that worksheets cannot be moved, deleted, hidden, unhidden, or renamed and new worksheets cannot be inserted, select the **Structure** check box.
(b) To use windows of the same size and position each time the workbook is opened, click the **Windows** check box.

As before, to prevent others from removing workbook protection, type a password, click OK, and then retype the password in the **Confirm Password** dialog box. Passwords are case sensitive – check that the **Caps Lock** key is not on.

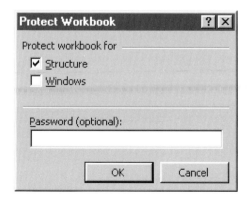

7.6 Removing protection from workbooks or worksheets[1]

If you are not sure whether you need to remove protection from a workbook or a worksheet to gain the access you want, go to **Tools Menu⇒Protection**. If the currently displayed worksheet is protected, the **Unprotect Sheet** command appears on the **Protection** menu. Similarly, if the workbook is protected, the **Unprotect Workbook** command appears. To determine whether a workbook is shared, look in the title bar for the word [**Shared**].

---

[1]   Note: In the Visual Basic Editor, programmers can use the EnableSelection property of Visual Basic for Applications to protect cells so that users cannot select them.

Chapter 8

Visual Basic

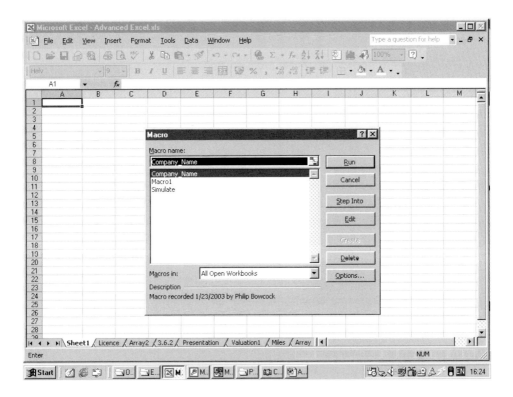

Microsoft has anticipated most of the general tasks one might require Excel to do. However there are still many specialist tasks that it cannot perform unaided. There are also some procedures that will be so specific to you, to your company or indeed to real-estate that it would be unrealistic for Microsoft to cater for all of these.

Visual Basic is a method of programming – the process of giving instructions to the computer to perform actions automatically. If, for example, you are assembling multi-tenanted cash flows, and using some of the tools considered in *Excel for Surveyors* such as Goal Seek, your instructions may be fairly complex. However anxious you may be to explore Visual Basic, check whether Excel can perform the task you need without programming. Even if you decide to programme a procedure, if there is an Excel provision the chances are this will run it more efficiently.

Visual Basic is a development of the BASIC (Beginner's All-Purpose Symbolic Instruction Code) programming language developed by John G Kemeny and Thomas E Kurtz at Dartmouth College, Hanover, NH, in the mid-1960s. This is one of the simplest high-level languages and has been learned with relative ease even by schoolchildren and novice programmers. It has re-appeared in various incarnations since then and this version, *Visual Basic for Applications*, was developed by Microsoft as part of the Office suite. It is also available as a stand-alone programming language, but in the context of Excel it can be used as an additional facility to enable routine and specialised calculations in Excel to be performed more efficiently, conveniently and automatically[1].

A general principle of programming is that if an operation is repetitive you should try to find a way for the computer do it for you. Visual Basic is the set of tools which help you write, edit and manage code to do this, and to offer assistance in the process. The best way to learn the process is to record a Macro, and then examine the code. Most programmers record as much as they can first and examine and modify the code afterwards rather than write a Macro from scratch.

As a final note although not difficult, VBA is a large and interesting subject and a glance at computer bookshelves will indicate the complexity which can be involved. Our objective here can only be to give an introduction to the subject and indicate possible uses to surveyors.

8.1 Macros

The starting point for discussion of Visual Basic is the Macro[2]. A Macro is a composite set of programming instructions which can be run by a single command such as pressing a key combination. In Excel this set of instructions can be created by making a recording of a series of actions which can then be 'played back'. It is essential to plan this recording to avoid including any unnecessary steps, and usually helpful to do a 'dry run' of the operations before starting.

Example: The following Macro records a procedure which will automatically enter your company name and phone number into any cell.

(a) Select an empty cell.
(b) Select **Tools Menu⇒Macro⇒Record New Macro**.
(c) A **Record Macro** dialog box appears. In the Macro name text box type: Company_name (note that spaces are not allowed and the Underline character must be used instead). Under **Shortcut-Key** type n. Then click OK.
(d) A **Stop Recording** toolbar will appear somewhere on your screen.
(e) Now carry out exactly those operations which you want to include in the Macro. Type the name of your company and your telephone number, eg

> Bayfield Training: 020 7866 8180

and click OK.

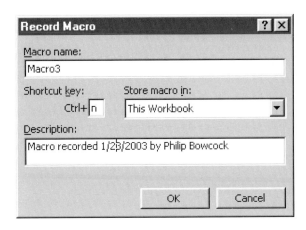

(f) Finally, to end recording, click the **Stop** button on the **Stop Recording** toolbar.

Now choose any empty cell and press **Ctrl** and n. This should result in the satisfying effect of automatically entering your company name and telephone number into this cell.

8.2 The Macro dialog box

The Macro dialog box allows you to mange and edit your Macros. Click on **Tools Menu⇒Macro** ... and **Macros**, or alternatively press Alt–F8.

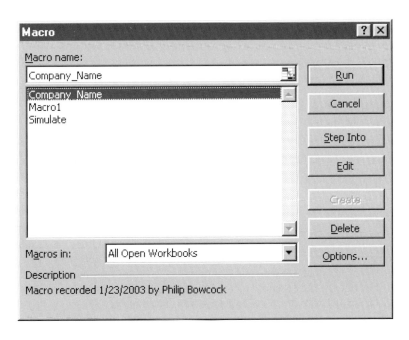

This box contains all the Macros available in either this workbook, all workbooks or your personal Macro workbook, depending on the choice you select in the box at the bottom. The personal Macro workbook is where we store Macros if they are to be available to all workbooks.

Four buttons with obvious uses are :

(a) Run: Selecting this dismisses the Macro dialog box and runs the Macro as if you had keyed the assigned shortcut.
(b) Cancel: Dismisses the Macro dialog box.
(c) Delete: Deletes the highlighted Macro
(d) Options: Takes you back into the Record Macro dialog box, allowing you to change the shortcut key and description.

8.2.2 Editing the Macro

Clicking on the **Edit** button will open Visual Basic alongside Excel on the task bar and we can flick between Excel and Visual Basic by clicking the appropriate button. Visual Basic now enables us to edit the code we created in the recording which should appear as follows :

```
Sub company_name()
 ' company_name Macro
 ' Macro recorded 5/8/2002 by Philip Bowcock
 ' Keyboard Shortcut: Ctrl+n
 ActiveCell.FormulaR1C1 = "Bayfield Training : 020 7866 8180"
 End Sub
```

The actions we recorded included both the machine code which carries out our steps and the corresponding text instructions. The machine code is not visible but using the text version we could add instructions which were not recorded, for example to allow for decision making, repetition of actions, or interaction with the Macro while it is running with user input forms.

There are three types of statements in a Macro :

(a) The Sub and End Sub statements mark the start and end the Macro and are essential. The Sub statement includes the name we gave to the Macro.
(b) Lines starting with an apostrophe are comments and are only there for the benefit of anyone examining it. It is often helpful to add additional comments as it is easy to forget precisely what each part of the code is actually doing, and it is good practice in any extended code to include such comments
(c) The code for the Macro called 'company-name' above consists of just one line. It is important to appreciate that Visual Basic is 'object oriented' which means that any line of code must start by specifying the object to be manipulated. The individual parts of the statement are as follows:

ActiveCell	The `ActiveCell` object refers to the cell that is currently active, and is followed by a full stop.
FormulaR1C1	A statement indicating that a simple assignment is to follow.
	The assignment operator, '=', has a meaning different from that normally attributed to this symbol ie equality. In Visual Basic, and most other types of code this means the 'value' on its right hand side is assigned, or given, to the 'object' on the left hand side.
="Bayfield Training: 020 7866 8180"	The text which is to be assigned to the active cell. Note that it is enclosed in quotes.

There are many objects available in Visual Basic. Every item you manipulate or use in Excel, for example a worksheet, a range of cells, or a chart has a name. This is why recording a Macro then viewing its code helps in learning the language.

An `ActiveCell` is an object because to Visual Basic it is something tangible that can be manipulated. Objects can be manipulated by changing their properties or methods. A property is a characteristic of an object, and a method is something the object does. Objects and their methods or properties are separated by a full stop.

In the above code `FormulaR1C1` is a property of the `ActiveCell` object. `FormulaR1C1` is very similar to another property called `Value`. In this case substituting `FormulaR1C1` with `Value` would produce the same result. The difference is what you actually key into a cell as opposed to the result. For example =2+2 is a formula, and 4 is a value.

We can now edit it by replacing the telephone number with our e-mail address. To do this we add to our code a message box object which is not available when simply recording a Macro.

Between the first line of code and End Sub add the following line:

Msg = 'To edit company-name Macro select **Tools Menu Macros** ...⇒**Edit**'

Notice how the text has been 'assigned' using '=' to the `MsgBox` object.

Leave Visual Basic open while you go back into Excel. Select Excel on the task bar by clicking the Excel icon in the top left hand corner of Visual Basic. Now select any empty cell and run your company-name Macro using the shortcut **Ctrl**+n to see the result of your amended code.

8.3 The VBA environment

The essential element of VBA is the machine code which cannot be seen. Everything else including the text version of the code is there to assist with managing, editing and using the code.

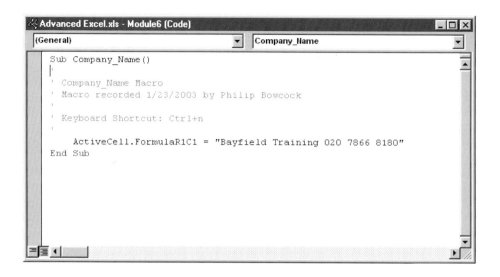

```
Advanced Excel.xls - Module6 (Code)                                    _ □ ×
(General)                            ▼   Company_Name                   ▼
  Sub Company_Name()
  |
  ' Company_Name Macro
  ' Macro recorded 1/23/2003 by Philip Bowcock
  '
  ' Keyboard Shortcut: Ctrl+n
  '
      ActiveCell.FormulaR1C1 = "Bayfield Training 020 7866 8180"
  End Sub
```

There are two toolbars, Standard and Edit, and three new buttons:

(a) The Excel button on the extreme left of the toolbar provides another way to switch back into Excel. (**Tools Menu⇒Macros⇒Visual Basic Editor** is the equivalent way to get from Excel back into Visual Basic.)

(b) The other two buttons to note on this toolbar are **Run** (**Play**) and **Stop**. So long as your cursor is positioned somewhere inside the code, the **Run** button allows you to run the code from Visual Basic. This is useful if you tile Excel and VBA side by side on your desktop. You can run the Macro, while keeping an eye on the code. If the code stops or does not work in Excel, it will be immediately obvious at exactly which line it has faltered. The line of the code in question will be marked by a bright yellow highlight. Note that if this occurs your code is running and has just paused at this line. It has not reached EndSub yet. To exit this situation and edit the code use the **Stop** button, represented by a square to the right of **Run**.

8.3.2 Windows

There are many windows providing different types of information, but the three most frequently used should appear by default.

The **Project Explorer** helps you navigate your code. Code in Visual Basic is contained in projects. Every time you open a workbook in Excel it creates a new project in Visual Basic. Each project contains objects. We have mentioned the `ActiveCell` object. This object only exists within another object: the familiar 'worksheet'. Also worksheets only exist within a 'workbook'. There are also fairly complex objects (have many properties and methods) that deserve distinction within the project. Examples of other complex objects that may be listed within a project are charts and forms.

When we record a Macro the code is stored in modules within the project. One module can contain many Macros. However we can add code to worksheets, forms and charts when we want to manipulate them specifically. For example we may want to embed a button into a specific worksheet.

To view the code of any item within the project double click on the object, and the code will appear in the code window normally on the right hand side of the screen.

The final window of importance is the properties window typically positioned at the bottom left of the screen. This allows you to review and set properties of various objects in your project without having to write code. This window is useful in more complex projects where larger objects such as worksheets are being manipulated.

8.4 The tidy_scenario macro

We used the scenario summary in *Excel for Surveyors* to provide us with a risk-adjusted return. In this example we will record a macro that, when run, will do some useful formatting for us. The purpose is to illustrate that learning Visual Basic is best done by recording macros, and examining the code.

Example: Produce a scenario summary with at least three results. Make sure one of the results is identical to the current model (see 14.3 of *Excel for Surveyors*).

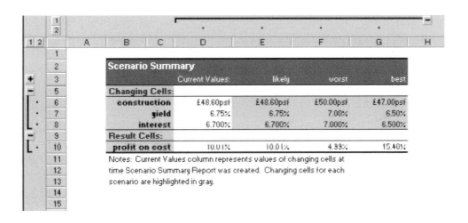

(a) Start a new macro and call it `tidy_scenario`.
(b) Select the entire sheet.
(c) Select **Data Menu⇒Group and Outline⇒Ungroup** (do this once for rows and once for columns).
(d) Select rows 3 to 5.
(e) Select **Format Menu⇒Row Unhide**.
(f) Select row 4, and **Edit Menu⇒Delete**.
(g) Select column D, and **Edit Menu⇒Delete**.
(h) Select **Tools Menu⇒Options⇒View** and check the gridlines box.

(i)   Select cell A1.
(j)   End recording.

This should produce the following results.

	A	B	C	D	E	F	G
1							
2		**Scenario Summary**					
3				likely	worst	best	
4		**Changing Cells:**					
5		construction		£48.60psf	£50.00psf	£47.00psf	
6		yield		6.75%	7.00%	6.50%	
7		interest		6.700%	7.000%	6.500%	
8		**Result Cells:**					
9		**profit on cost**		10.01%	4.99%	15.46%	
10		Notes: Current Values column represents values of changing cells at					
11		time Scenario Summary Report was created. Changing cells for each					
12		scenario are highlighted in gray.					
13							

Delete this sheet and produce another one by selecting the summary button in the scenario manager again. Then run your tidy_scenario macro, and watch Excel do all the tedious work for you.

The text at the bottom will be in a different place each time depending on the size of the scenario.

View the code by selecting **Macro Menu⇒Macros**, and **Edit** when you should see the following code.

```
Sub tidy_scenario()
'
' tidy_macro Macro
' Macro recorded 10/2/2002 by Natalie Bayfield
'
Cells.Select
Selection.Rows.Ungroup
Selection.Columns.Ungroup
Rows('3:5').Select
Selection.EntireRow.Hidden = False
Rows('4:4').Select
Selection.Delete Shift:=xlUp
Columns('D:D').Select
Selection.Delete Shift:=xlToLeft
ActiveWindow.DisplayGridlines = True
Range('A1').Select
End Sub
```

Read through the code, and see how similar the words are to the instructions we gave for recording the macro. Code follows instructions in sequence and each line works because of the previous lines of code. For example the line of code 'Rows('4:4').Select' selects row 4. Therefore the next line of code 'Selection.Delete' knows which line to delete. However we can consolidate the code by combining statements:

```
Sub tidy_macro()
'
' tidy_macro Macro
' Macro recorded 10/2/2001 by Natalie Bayfield
'
Rows.Ungroup
Columns.Ungroup
Rows('3:5').EntireRow.Hidden = False
Rows('4:4').Delete Shift:=xlUp
Columns('D:D').Delete Shift:=xlToLeft
ActiveWindow.DisplayGridlines = True
Range('A1').Select
End Sub
```

Delete the scenario summary sheet, and produce another from the scenario manager in the original sheet. Run the macro on this new summary again. This time you will notice that the sheet, rows, or columns are not selected throughout the execution of the code.

However there still appear to be a couple of jumps in the spreadsheet when you run the code. This is caused by the visual elements Shift:xlUp, and Shift:xlToLeft to the right of the Row and Column delete lines of code. Since these are visual element they can be removed without affecting the end result. Run the code again on yet a new summary using the edited code. You will notice now how much cleaner it runs.

8.5 Stepping through code

Example: After editing a complicated Macro it is quite likely that will not do exactly as you expected and one means to isolate the problem is to step through the code. This way you can watch the Macro execute line by line.

(a) First ensure that the code window is selected, and the `tidy_scenario` Macro is in view.
(b) Close the **Project Explorer** and **Properties** windows by clicking the close box on the top right corner as usual, and then maximise the code window if it hasn't already done so automatically.
(c) The Excel icon should be next to the Visual Basic icon on the task bar. Right mouse click on the task bar and select **Tile Vertically**. This will place Visual Basic and Excel side by side. If you have other programs open minimise these.
(d) In Excel create a new summary sheet.

(e) Click anywhere within the code and press F8 on the keyboard. This will highlight the first line of code in yellow.

(f) Keep pressing F8 and watch it execute the code in Excel.

Each time you press F8 it will execute the highlighted line, and then move onto the next line, highlight it and stop. This procedure enables you to check the code as it operates, and find any errors.

8.6 Conclusion

After reading this chapter you may think of many possible uses of Visual Basic such as extending cash flows automatically, collapsing cash flows between annual and quarterly, adding tenants to lists, and so on. The possibilities are quite literally endless. You could even build your own automatic valuation program from scratch (although for reasons we have discussed this is not to be recommended).

As we have indicated previously, very complicated Macros can be created and for further information we set out some references in Appendix A.

Chapter 9

Regression analysis

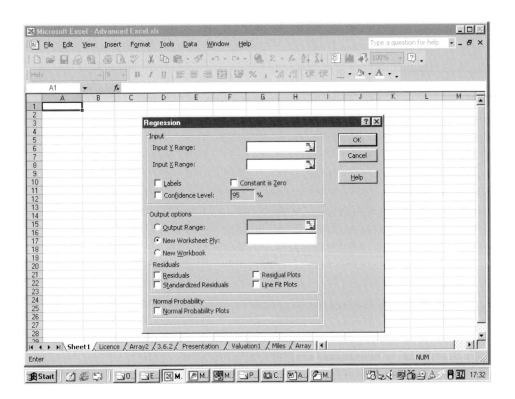

9.1 Introduction

This is an analysis tool which can be used to determine the relevance and significance of the various elements which, taken together, contribute to the value of a property or a yield. In time immemorial (ie before 1952 when the first computer appeared) the analysis of data was for practical purposes limited to work which could be done with paper and pencil, with possibly an adding machine and a slide rule. Since that time the ever-increasing availability of computing power has enabled analysis to be done which would previously been far too laborious, and in many cases impossible. Of course there is no point in analysing property data (or any other data for that matter) unless there is a reasonable prospect that the results will useful in practice.

There are many possible uses for the results of a regression analysis, for example the establishment of a level of values in an area where a major change has taken place (the effect of flooding on the value of houses) or a decision on where to site a new supermarket.

Economic theory traditionally sets out a set of relationships between variable factors, such as supply and demand, and demonstrates that there is a relationship between them, for example, that the higher the price the less will be demanded and the more will be offered for sale. Textbooks on economic theory display multitudes of graphs of these relationships and first-year students of economics can be forgiven for assuming that it is impossible to make precise calculations about prices, profits or elasticities of demand and supply.

The actual measurement of economic factors is in the province of Econometrics which is broadly a discipline incorporating elements of economics, mathematics and statistics. Multiple regression analysis is one technique among the general class of Multivariate Analysis, and is one we can use to analyse the effect of each independent variable on the dependent variable.

9.2 The property market

The practice of valuation – estimating the amount of money for which a property might be expected to sell or the value to an individual of an asset which he owns or proposes to acquire – is fundamentally dependent on the analysis of other transactions, market trends, and professional opinions.

When it comes to the real world the problem is by no means as clear cut. For example in the particular case of the residential property market each house is different – physically identical houses on the same estate may have different outlooks. Even in the case of new properties, each house stands on its own unique site, and after a few years houses which were originally structurally identical will have different attributes – extensions, style of decorations, condition of garden, structural and decorative repair, etc. Sale data is not always consistent or reliable – in some instances it happens that the vendor of a house will sell to a purchaser at a price less than he could obtain from another bidder simply because he likes the look of the purchaser and prefers to hand over the house in which he has lived for many years to someone who, he considers, will take care of it as he would have done. Conversely, a particular purchaser may be prepared to pay more for a property for a special personal or taxation advantage.

The object of regression analysis in the property context is to take a set of observations of transactions and data about the subject properties, and from these endeavour to ascertain by analysis the most important factors determining the price. We might consider that the size of the property is an important factor but the questions remain – how important is this, and what are the other factors contributing to the value?

9.3 How not to do it!

A simple analysis of residential property might proceed as follows, based on the following data :

No	Type	Area	Garage	Sale Price
1	Semi	800	0	80,000
2	Semi	900	1	100,000
3	Detached	1000	1	130,000

From the above data one might be tempted to conclude the following :

Price per square foot for basic house	£100
Extra for garage	£10,000
Extra for being detached	£20,000

No valuer today would regard this as a satisfactory approach to the analysis of property values because the sample is far too small, (although in the past it has been done like this in practice!). This approach would become impossible if two identical houses were sold for different prices, quite a common situation in the market. However if we extended the sample to 100 properties and again looked for differences which could be analysed in this manner we would have to find 100 characteristics which could differentiate between properties, and in fact the situation becomes worse for two reasons.

First, in mathematical terms we are dealing with a series of simultaneous equations, and small changes in input data can very often produce wildly differing results (try for example changing the area of House No 1 above to 805 square feet and see the difference). Secondly, the only practical way of finding the solution to 100 simultaneous equations would be by computer, and it is very likely that substantial rounding errors would occur, even with the usual computer precision of 15 significant figures.

9.4 The alternative approach

The object of regression analysis is to abandon absolute precision and look for a more limited set of characteristics (variables) which (a) would explain most of the variation in prices, and (b) at the same time be sufficiently robust as not to be affected by small errors in the data.

The method is frequently referred to as 'least squares'. We can consider this concept by analogy with Pythagoras' theorem. In the triangle above the hypotenuse AB represents all the variables which would explain the values of every property, the base BC represents the explanation derived from a limited number of variables and the perpendicular AC represents the variation not explained by the model. Clearly the best explanation or 'best fit' is obtained when AC is perpendicular to BC. (This is an example of the reason why so many statistical formulae involve squares and square roots of numbers.)

9.5 Terminology

Before proceeding further, we define some of the terms we shall use:

(a) Dependent variable. The variable we are investigating, in this case the price of houses. The price is dependent on the other data.

(b) Independent variable. One or more variables which have an effect on the dependent variable. In this example there is only one independent variable – the area.

(c) The trend line. This is often referred to as the 'line of best fit' or 'regression line'.

(d) Residual. In each case the difference between the line and the actual data point.

9.6 The data

The first part of the procedure is to obtain data and to verify its accuracy as far as possible. It is a fundamental characteristic of this that data is more easily counted than measured: it is easier to be precise about the number of rooms in a house than to be precise about the area of one room. The maxim of computer operations of 'garbage in – garbage out' applies with full force in regression analysis.

Besides checking the actual data it is important to ensure that it is relevant to the analysis. For example it would clearly be inappropriate to include in a sample of housing data a property with a large garden which could have development value, or a £multi-million mansion.

A simple regression is frequently used as a starting point for analysis in order to establish the relationship between just two variables. In the example which follows we assume data of property prices in a typical town, and begin by plotting these against the internal areas and other features we shall consider later. It is apparent from the plot that larger houses tend to be worth more than smaller houses, but there is a wide variation and so further consideration is appropriate. We need a means of drawing more specific conclusions about the data if our work is to be useful.

9.7 Simple linear regression

Below is the data for the multiple regression example which we shall use later. We only used Price and Area data in the following example.

Example Data

No	Price	Detached	Terrace	Floors	Area sq.ft	Cloak-room	Garage
1	60950	0	1	0	591	0	0
2	62950	0	1	0	507	0	0
3	65000	0	1	0	599	0	0
4	69950	0	1	0	681	0	0
5	69950	0	0	0	353	0	0
6	72950	0	1	0	743	0	0
7	76000	0	1	0	577	1	1
8	82500	0	0	0	601	1	0
9	84950	0	0	0	683	0	0
10	85000	0	1	0	752	0	1

No	Price	Detached	Terrace	Floors	Area sq.ft	Cloak-room	Garage
11	86950	0	1	0	784	1	1
12	88500	0	1	0	981	0	0
13	88950	0	1	0	624	0	1
14	89950	0	0	0	724	0	0
15	90000	1	0	0	683	0	0
16	92000	0	1	0	995	1	0
17	95000	1	0	1	600	0	0
18	95000	0	0	1	591	1	1
19	95000	0	1	0	757	0	1
20	95000	0	0	0	587	0	1
21	96000	0	0	0	725	1	1
22	96500	0	0	0	679	1	1
23	97950	1	0	1	574	0	0
24	99950	0	0	0	694	1	1
25	99950	0	0	0	711	0	1
26	99950	0	0	0	753	1	1
27	102950	0	0	1	550	1	1
28	102950	0	0	0	662	1	1
29	104950	0	0	0	754	0	1
30	105000	0	0	0	651	1	1
31	105000	0	0	0	830	1	1
32	105950	0	0	0	858	0	1
33	105950	0	1	0	1063	1	0
34	107950	0	0	0	740	1	1
35	109950	1	0	1	778	1	1
36	110000	0	0	0	1036	0	1
37	110000	1	0	0	787	1	1
38	112000	0	0	0	898	1	1
39	113500	1	0	0	714	1	1
40	116500	0	0	0	898	1	1
41	116950	1	0	1	746	1	1
42	119500	1	0	0	747	0	1
43	119950	1	0	0	639	0	1
44	119950	1	0	1	749	0	1
45	119950	1	0	0	701	0	1
46	119950	0	0	0	1111	0	1
47	120000	1	0	0	891	1	1
48	120000	1	0	0	477	1	1
49	120000	0	0	0	881	1	1
50	120000	1	0	0	685	0	1
51	121000	1	0	1	784	0	1
52	122950	1	0	1	914	0	1
53	125000	1	0	0	1045	1	1
54	125000	1	0	0	892	1	1
55	125000	1	0	0	986	1	1
56	129950	1	0	0	994	0	1
57	130000	1	0	0	797	1	1
58	130000	1	0	1	1029	1	1
59	130000	1	0	0	906	1	1

No	Price	Detached	Terrace	Floors	Area sq.ft	Cloak-room	Garage
60	130000	1	0	1	1130	1	1
61	133000	1	0	0	1174	1	1
62	133950	1	0	0	921	1	1
63	135000	1	0	1	925	0	1
64	135000	1	0	0	974	1	1
65	135000	1	0	0	1270	1	1
66	137950	1	0	1	1024	0	1
67	139950	1	0	1	844	0	1
68	140000	1	0	0	1267	0	1
69	140950	1	0	1	1055	1	1
70	145995	1	0	1	1094	0	1
71	160000	1	0	0	1481	1	1
72	162950	1	0	0	1400	1	0
73	195000	1	0	0	2461	1	1

9.8 Running a simple linear regression analysis using Excel

We now select the regression analysis add-in which is found in **Tools Menu**⇒**Data Analysis** ... (If this is not present see Appendix 3.)

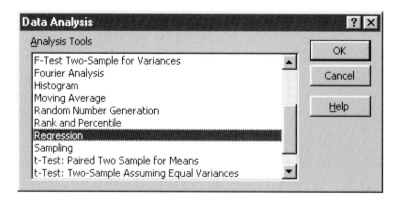

In the **Data Analysis** dialogue box select **Regression** and the next dialogue box will appear. Enter details as shown. The **Input Y Range** is the column of values, and the **Input X Range** is the column of areas, including the column headers which become the labels. Click the Labels box so that these are not treated as part of the data.

Enter a cell reference as the output range so that results will appear on the same worksheet. Choose a cell to the right of the data to mark the top left cell of the output data which will include several cells.

Click on **Residuals, Residual Plots** so that you can compare the results with the original data, and produce a chart of the results.

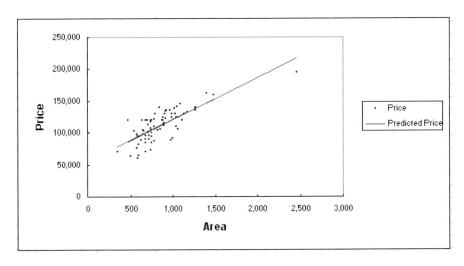

Regression

Input

Input Y Range: `$B$1:$B$74`

Input X Range: `$F$1:$F$74`

☑ Labels ☐ Constant is Zero

☐ Confidence Level: `95` %

Output options

◉ Output Range: `$J$1`

○ New Worksheet Ply:

○ New Workbook

Residuals

☐ Residuals ☐ Residual Plots

☐ Standardized Residuals ☐ Line Fit Plots

Normal Probability

☐ Normal Probability Plots

OK

Cancel

Help

Click OK and the analysis will appear alongside the data.

The chart below shows price plotted against area and the upward sloping Trend Line. One could of course estimate this by careful plotting on paper and laying a ruler over it. However statistical theory is more rigorous, and by use of the formula determines the line such that the sum of the squares of the distances from each plotted point to the line is minimised. (The formatting of the chart will probably need some adjustment.)

Numerous statistics are also produced:

J	K	L	M	N	O	P	Q	R
SUMMARY OUTPUT								
Regression Statistics								
Multiple R	0.774807							
R Square	0.600325							
Adjusted R Square	0.594696							
Standard Error	15575.81							
Observations	73							
ANOVA								
	df	*SS*	*MS*	*F*	*Significance F*			
Regression	1	2.59E+10	2.59E+10	106.6445	8.76E-16			
Residual	71	1.72E+10	2.43E+08					
Total	72	4.31E+10						
	Coefficients	*Standard Error*	*t Stat*	*P-value*	*Lower 95%*	*Upper 95%*	*Lower 95.0%*	*Upper 95.0%*
Intercept	55415.34	5690.896	9.737541	1.03E-14	44068.01	66762.67	44068.01	66762.67
Area	65.80349	6.372059	10.32688	8.76E-16	53.09796	78.50902	53.09796	78.50902

The most significant of these are:

(a) R Square. This shows the proportion of the variation in Price which is explained by the Area data alone. In this case we have only explained 60% – there are obviously more factors influencing price. A value of zero indicates that nothing has been explained while a value of 1 indicates that all the variation has been explained. Beware however that a very high value (close to 1.0) may be suspect.

(b) The Standard Error. This is the standard deviation of the differences between the actual and predicted values. At £15,575 this will not be very helpful in any predictions.

(c) The Coefficients. The Intercept is the constant b in the formula y = ax + b which is the formula for the regression line. The area coefficient is the price per square foot to be added to the constant to give the predicted value.

A discussion of the remaining statistics can be found in any substantial book on the subject, and will not be considered here.

9.9 Multiple linear regression

It is clear that using only Area as an independent variable fails to explain much of the variation in price and that other factors are significant. We therefore turn to Multiple Regression Analysis so that these can be considered.

We now use all the data from the previous example to see to what extent the explanation of prices can be improved. In this case our dependent variable is Price and independent variables are Detached, Terrace, Single floor (ie Bungalow), Area, Cloakroom and Garage. (Excel can deal with a maximum of 16 variables.)

9.10 Types of variables

As we have seen, an independent variable is a factor which may affect the value of the set of properties – whether it actually does or not is one of the objects to establish. The dependent variable is the one which we are investigating, possibly with the intention of using it for prediction.

9.10.1 Metric variables

As its name implies, this is a variable which can be measured, such as an area of a property or the population of a town. Obviously this should be done accurately, but accuracy is often a relative matter, and there is little point in spending a great deal of effort in measuring one variable to high precision when another can only be done with some approximation. The variable of area in the previous example was clearly a metric variable.

9.10.2 Dummy variables

Some variables, such as whether or not a house is detached cannot be measured – it is either detached or not as the case may be, but it is nevertheless likely to be a significant factor in the determination of the value. This variable therefore has a value of zero (not detached) or 1 (detached), and is known as a dummy variable.

9.11 Correlated variables

Frequently two variables have the same values for each observation. For example, using houses again, if every detached house has a second garage and no other house has one, allocating value between 'detached' and 'second garage' will be impossible. There is no way that the computer could distinguish between them, and the computer will display an error. This is an example of perfect correlation between two independent variables.

However it is possible for two variables to be nearly correlated, and the problem here is that although a solution is possible, small variations in the data relating to these variables can produce substantially different results.

Therefore it is necessary to use the Correlation Matrix to examine the data for the possibility of correlation.

The Excel procedure for calculating the correlation matrix is similar to that of the regression analysis itself. Select **Tools Menu⇒Data Analysis** ... and then **Correlation**. This box will request the range of data required – drag over the data set including the header row (maximum 16 variables) and click the Labels box. Also click on a cell to the right of the data which will set the top corner cell for the results.

Click OK and the correlation matrix will be calculated.

This matrix shows the correlations between each pair of variables.

The object now is to check the matrix for values which indicate a close correlation between two variables. If 1.000 appears anywhere (except the leading diagonal which is always 1.000) this indicates perfect correlation and the regression will not run if both variables are included (obviously it would be

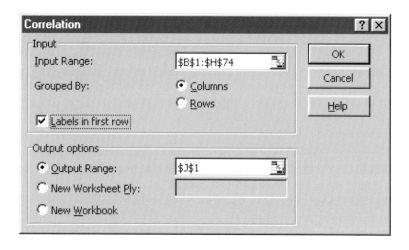

J	K	L	M	N	O	P	Q
	Price	Detached	Terrace	Floors	Area	Cloakroom	Garage
Price	1.00000						
Detached	0.70604	1.00000					
Terrace	-0.58059	-0.47190	1.00000				
Floors	0.22370	0.39014	-0.24661	1.00000			
Area	0.77481	0.36716	-0.16836	-0.01736	1.00000		
Cloakroom	0.28480	0.06772	-0.21141	-0.10276	0.23905	1.00000	
Garage	0.52591	0.27219	-0.44580	0.12061	0.20352	0.30190	1.0000

impossible to distinguish between the effects of two variables which are perfectly correlated). Delete one, and remember that the remaining one will include the effect of both.

If one of the cells contains a value approaching 1.000 this will indicate a close correlation which will not necessarily be fatal, but can distort the results. As a rule of thumb consider that if the value is greater than 0.900, one of the variables should be removed.

In our case neither of these problems occur, so we can proceed with the regression.

9.12 Running the regression

Use the same approach as with the single regression, but include all the independent variables, as with the correlation matrix. The following statistics appear.

We have now substantially improved on the single regression. The R Square is 0.923, which is typical for regressions of such properties. The standard error of £7,048 is also a typical statistic. Below, the Coefficients column gives the value for each variable, for instance the Area shows £48.67 per square foot with a standard error of £3.24. If the property is detached it will be worth an additional predicted £15,411, with a standard error of £2,133.

SUMMARY OUTPUT								
Regression Statistics								
Multiple R	0.96121							
R Square	0.92393							
Adjusted R S	0.91701							
Standard Err	7048							
Observation:	73							
ANOVA								
	df	*SS*	*MS*	*F*	*Significance F*			
Regression	6	3.9819E+10	6636489817	133.5949	6.149E-35			
Residual	66	3278630902	49676226					
Total	72	4.3098E+10						
	Coefficients	*Standard Err*	*t Stat*	*P-value*	*Lower 95%*	*Upper 95%*	*Lower 95.0%*	*Upper 95.0%*
Intercept	54133.88	3206.79	16.8810	0.00000	47731.31	60536.45	47731.31	60536.45
Detached	15411.11	2133.93	7.2219	0.00000	11150.58	19671.63	11150.58	19671.63
Terrace	-14543.45	2669.07	-5.4489	0.00000	-19872.42	-9214.49	-19872.42	-9214.49
Floors	1738.60	2229.47	0.7798	0.43828	-2712.69	6189.88	-2712.69	6189.88
Area	48.67	3.24	15.0198	0.00000	42.20	55.14	42.20	55.14
Cloakroom	822.41	1800.65	0.4567	0.64937	-2772.70	4417.51	-2772.70	4417.51
Garage	12463.66	2310.58	5.3942	0.00000	7850.43	17076.89	7850.43	17076.89

(*Tip*: most of these statistics can be examined more easily if you format the cells to a smaller number of decimal places, say, for most, 4 places. Some, such as the predicted values, are more conveniently shown as zero decimal places.)

From these coefficients the value of each property can be calculated and part of this table is shown below. For each case, the value predicted by the equation and the difference between this and the actual price (the Residual) are shown.

Many other statistics are also calculated, and these can be used to further refine the equation by removing variables which are statistically unsound, but a consideration of these is beyond the scope of this book. Consult any good text on statistics for an explanation.

9.13 Residual plot

Finally, it is possible to plot the results of the analysis and to consider whether any of the observations should be considered as outliers and excluded from a second run. Use the Chart Wizard and select the residuals to produce a scatter plot. Observations with large residuals should be examined to consider whether it is appropriate to include them in the analysis.

Below are the first few residuals.

From these residuals we can plot another chart that we can use to consider whether any of the observations should be rejected as being untypical of the market. For example No 73 has a residual of nearly £25,000, and on looking at the original data we see that it is much larger and the price is greater than any of the others. A more consistent result would almost certainly have been obtained if this one had been omitted, and a re-run without this would be worth considering.

	I	J	K	L
26				
27		RESIDUAL OUTPUT		
28				
29		*Observation*	*Predicted Pric*	*Residuals*
30		1	68376.34	-7426.3425
31		2	64253.64	-1303.6393
32		3	68756	-3756.0012
33		4	72722.95	-2772.947
34		5	71296.4	-1346.3987
35		6	75765.08	-2815.0835
36		7	80966.37	-4966.3651
37		8	84204.6	-1704.6041
38		9	87383.22	-2433.216
39		10	88657.07	-3657.0737
40		11	91017.58	-4067.5838
41		12	87320.33	1179.66558
42		13	82421.91	6528.08914
43		14	89383.72	566.275072
44		15	102808.9	-12808.924
45		16	88838.78	3161.21883
46		17	100488.1	-5488.0941
47		18	97929.85	-2929.8519
48		19	88910.18	6089.82059
49		20	95159.55	-159.54962
50		21	102694.1	-6694.1253
51		22	100474.6	-3974.5826
52		23	99232.3	-1282.3002
53		24	101219.3	-1269.2976

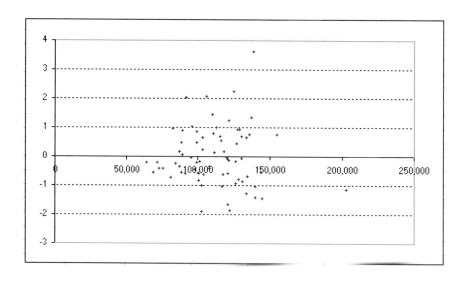

Chapter 10

Linear programming

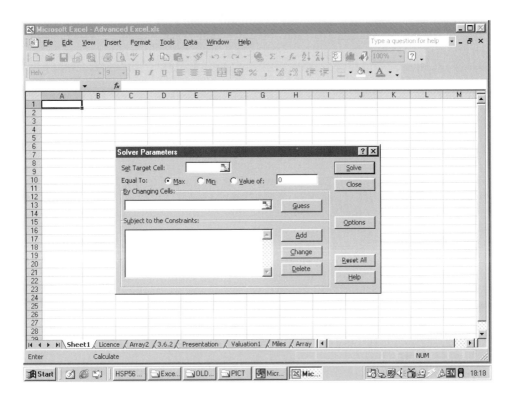

10.1 The problem

In some situations we may need to find the optimum solution to a problem involving inputs which are subject to various constraints, for example the optimum number and type of dwellings to be erected in a development, or the optimum method of investment where there are numerous possibilities and constraints on capital.

The technique of linear programming which has been used for many years in other disciplines is relevant to decisions such as this, and can be developed with the use of Excel.

As a simple example we take the problem of a residential development scheme. The normal objective of the developing company is to maximise profitability by building the largest number of houses at the lowest cost per house and selling at the highest price. In practice of course it is not as simple as that, and there are always constraints on the project. The first obvious one is the size of the site which cannot be changed. Then there will be restrictions imposed by the planning authority requiring a minimum number of units; a proportion for low-income purchasers; the results of market research which perhaps indicates a demand for large 'executive' houses on large sites; and projections of likely sales prices on completion for the different types.

As a very simple example we first consider a site where the developer can physically build 100 detached houses (D) or 80 pairs of semi-detached (S) or any linear combination of these two. Net profits on houses are expected to be £6,800 (D) and £8,000 (S). Market research indicates that there would be a maximum demand for 75 (S) but the planning authority would require a minimum of 30 (S). Market research also indicates that there would not be a demand for more than 45 (D). How many of each type should be built?

10.2 A graphical solution

The problem can now be expressed in graphical form as shown. In the chart below the x-axis represents the number of detached houses that can be built and the y-axis the number of semi-detached pairs. The various combinations which can physically be constructed are shown on the diagonal line AA. Lines show the maximum number of detached houses and the maximum and minimum numbers of semi-detached. The shaded area depicts all the combinations of the two types ('the feasible area') which can be built subject to the above constraints.

We now consider profitability. The profit from 85 semi-detached pairs will equal the profit from 100 detached, and clearly any linear combination of the two types will produce the same amount of profit. This is shown by the line PP on our chart. However all lines parallel to this will show the profit from proportionate increases in numbers, and the greatest profit will be where one of the parallel lines touches point X. The greatest profit is therefore obtained by building 24 detached and 70 pairs of semi-detached.

If the relative profits from the two types of house change the profit slope (PP) might well become steeper in which case the most profitable combination could become 45 detached and 30 semis (point Y).

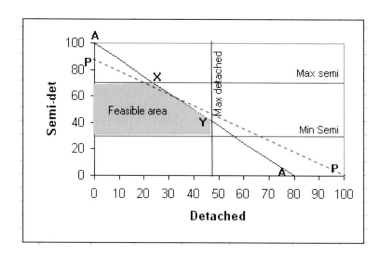

Obviously this is a very simple problem and is unlikely to require the advanced techniques of linear programming. If there were three types of house in our example a three-dimensional model would be required, and beyond that it becomes impossible to create a graphical description. However multi-dimensional problems can still be handled in mathematical terms, and Excel's Solver function has the facilities to deal with these. You may like to try the above problem using Solver as we now describe.

10.3 Solver

We briefly mentioned the solver function in *Excel for Surveyors*, section 15.4. Solver is similar to the Goal Seek function in that we specify an answer and ask the computer to keep trying the inputs until it gets close to that answer. However Goal Seek is fairly limited since we have to specify the desired result and can only change one variable.

Solver is much more powerful and we can use this function to calculate the optimum solution where there are many variables and several constraints, as in the following example.

Example: A site of 5 hectares (50,000 m²) is to be developed and details of types, costs and prices are set out below:

Type	Beds	m^2	Cost/m^2	Plot size m^2	Cost/ unit	Sale/unit
Detached	5	275	500	1,000	137,500	400,000
Detached	4	250	520	700	130,000	350,000
Detached	3	220	540	500	118,800	300,000
Semi-det	3	200	550	400	110,000	200,000
Semi-det	2	150	500	300	75,000	120,000
Maisonette	2	120	600	250	72,000	100,000
Maisonette	1	100	550	200	55,000	60,000

The following constraints must be taken into account.

(a) Obviously numbers of units must be integers. You cannot build one third of a semi and two-thirds of a detached house.
(b) Similarly semi-detached houses must obviously be built in pairs.
(c) The maximum size of the site is 50,000 m².
(d) We obviously do not want to make a loss on any type of house.

We will also assume that the local planning authority requires a minimum of 150 units, and that at least 20% should be 'affordable'.

We first set up a worksheet containing all the information above and the calculations which will indicate the profit on completion. The number of houses etc is set at zero to start.

Columns A, B, D, E, F, H and J contain the information already given. Column C will eventually contain the numbers calculated by the computer, and columns G, I, K and L will be the calculated results of the proposed numbers. Cell L9 is the profit from the development and is the amount we set out to maximise.

	A	B	C	D	E	F	G	H	I	J	K	L
1	Type	Beds	No	m2	Cost /m2	Plot - m2	Total of plots	Cost /unit	Total cost	Sale /unit	Total sale	Profit
2	Det	5	0	275	500	1,000	0	137,500	0	400,000	0	0
3	Det	4	0	250	520	700	0	130,000	0	350,000	0	0
4	Det	3	0	220	540	500	0	118,800	0	300,000	0	0
5	Semi	3	0	200	550	400	0	110,000	0	200,000	0	0
6	Semi	2	0	150	500	300	0	75,000	0	120,000	0	0
7	M'ette	2	0	120	600	250	0	72,000	0	100,000	0	0
8	M'ette	1	0	100	550	200	0	55,000	0	60,000	0	0

10.4 Setting the constraints

Specifying the constraints must be done with care. Note that it is only possible to specify one constraint at a time, so that in the case of Cells C2:C8 we must specify that they are to be equal to or greater than zero, and also they must be integers.

Select **Tools Menu**⇒**Solver** to generate the following window.

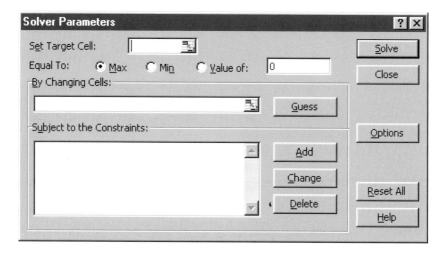

Enter details as follows:

The target cell (the profit to be maximised) is L9. Ensure that the 'Max' button is clicked.

The cells to be changed (the numbers of units of each type to be built) are C2:C8 To specify the constraints click on the 'Add' button and the next dialogue appears:

For cell C2 we shall need the following constraints:

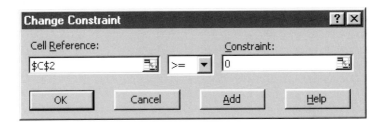

Click on OK and each constraint will be added to the list. C9 must be set to a minimum of 150 (the number required by the local planning authority) and G9 to a maximum of 50,000 (the total size of the site). L2:L8 are each set to a minimum of zero to ensure that each type is profitable. The complete list of objectives now appears as:

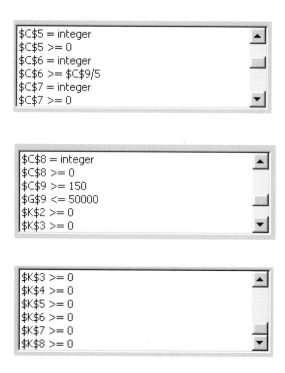

```
$C$5 = integer
$C$5 >= 0
$C$6 = integer
$C$6 >= $C$9/5
$C$7 = integer
$C$7 >= 0
```

```
$C$8 = integer
$C$8 >= 0
$C$9 >= 150
$G$9 <= 50000
$K$2 >= 0
$K$3 >= 0
```

```
$K$3 >= 0
$K$4 >= 0
$K$5 >= 0
$K$6 >= 0
$K$7 >= 0
$K$8 >= 0
```

Before clicking on 'Solve' check the Options button which should show the following dialogue. If not, change it accordingly. There are several other options available which are not examined here – try them for yourself to see the effect.

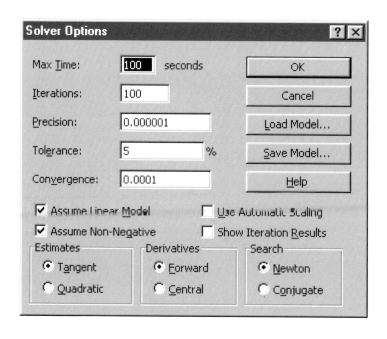

Solver Options ? X

Max Time:	100 seconds	OK
Iterations:	100	Cancel
Precision:	0.000001	Load Model...
Tolerance:	5 %	Save Model...
Convergence:	0.0001	Help

☑ Assume Linear Model   ☐ Use Automatic Scaling
☑ Assume Non-Negative   ☐ Show Iteration Results

Estimates
 ● Tangent
 ○ Quadratic

Derivatives
 ● Forward
 ○ Central

Search
 ● Newton
 ○ Conjugate

Now click OK and then Solve to see the results:

	A	B	C	D	E	F	G	H	I	J	K	L
1	Type	Beds	No	m2	Cost /m2	Plot - m2	Total of plots	Cost /unit	Total cost	Sale /unit	Total sale	Profit
2	Det	5	0	275	500	1,000	0	137,500	0	400,000	0	0
3	Det	4	0	250	520	700	0	130,000	0	350,000	0	0
4	Det	3	56	220	540	500	28,000	118,800	6,652,800	300,000	16,800,000	10,147,200
5	Semi	3	0	200	550	400	0	110,000	0	200,000	0	0
6	Semi	2	32	150	500	300	9,600	75,000	2,400,000	120,000	3,840,000	1,440,000
7	M'ette	2	0	120	600	250	0	72,000	0	100,000	0	0
8	M'ette	1	62	100	550	200	12,400	55,000	3,410,000	60,000	3,720,000	310,000
9			150				50,000					11,897,200
10												

Solver has found that the maximum profit is obtained by building 56 3-bedroom detached, 32 2-bedroom semi-detached and 62 1-bedroom maisonettes.

Clearly, in an actual scheme there would be other factors to consider such as the practicalities of the site layout and building regulations, but a linear programming approach can be helpful in initially formulating a scheme. In a larger scheme which is developed over a period of years it may be appropriate to re-run Solver from time to time with updated information to ascertain whether the mix of units should be revised.

Chapter 11

Simulation

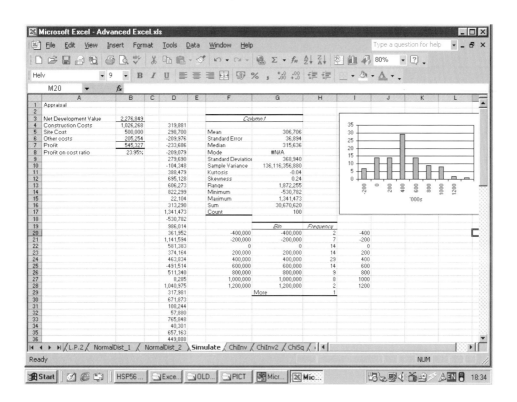

11.1 A little statistics

11.1.1 The Mean or Average

We considered the Average function in Chapter 5 of *Excel for Surveyors*, although at the time this was simply to demonstrate the use of functions in Excel. In Chapter 14.3 we also discussed the Scenario Manager which allows several valuations based on different assumptions to be compared. Most surveyors in practice will usually make several draft appraisals using different rates of interest, rental values, assumed growth rates, etc, before selecting a final opinion. The Scenario Manager is Excel's facility for using the computer to do this.

One approach to a final decision on the value is simply to take the mean of the various answers obtained. Another is to use the Median, which is the mid-point of answers, ie the one 'in the middle'. Neither of these tell us anything about the spread of the different answers so far. This is where the Standard Deviation comes in.

11.1.2 The Standard Deviation

If we were to do, say, 10, valuations with different assumptions about rents, rates of interest, outgoings, etc, we might expect a tendency for the answers to clump together towards the mean, but with some outliers. This 'spread' of the outliers is measured by the Standard Deviation.

To calculate this for a given set of data we proceed as follows:

(a)   Calculate the mean.
(b)   For each answer ('observation') calculate the difference between the mean and the observation and square it.
(c)   Sum these differences.
(d)   Take the square root of the total. This is the Standard Deviation.

Obviously this would be very laborious by hand but fortunately there are many computer programs and even pocket calculators which will carry out the operation for us. Excel of course has this facility.

11.1.3 Expectation

In valuation terms the statistical expectation of a calculation may be described as the value multiplied by the probability of getting it. In Chapter 14.5 of *Excel for Surveyors* we demonstrated a simple risk model in which we multiplied each valuation by the probability of getting it and summed the results. This is a much more subjective and less reliable method than applying probabilities to the input variables instead. This is because each appraisal is a unique mix of input variables each with its own pattern of risk. Applying probabilities to the results means that the same probabilities are being applied to all the input variables as well.

11.1.4 Distributions

The probabilities applied to each of the outcomes in the previous book is more appropriately described as a prediction rather than a distribution. A distribution is a description of the possible values which a variable can take. For an example we need look no further than decisions of the courts in negligence claims, where expert witnesses can produce numerous different answers to the question, 'What is the property worth?'.

11.1.5 The Normal Distribution

Statisticians have developed many theoretical distributions of data, but one of the most commonly used is the Normal Distribution. A full description and specification of this can be found in most books on statistics, but for the present we will simply describe it as having a symmetrical bell-shaped graph. The bell shape indicates that most observations of a set of data will be close to the mean. The further an observation is from the mean the less likely it is to occur. The symmetry indicates that the mean is also the median.

A particular case of the Normal Distribution is the Standard Normal Distribution, which has a mean of zero and a standard deviation of 1. Its graph is:

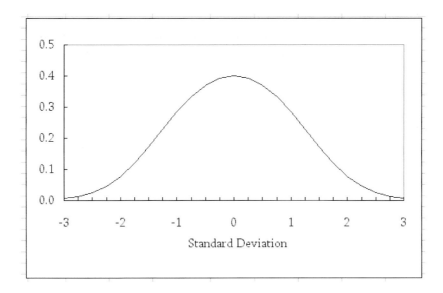

For example, in a valuation we might expect that the most likely rate of interest is 7% but that 7.25% is fairly likely, 7.5% a little less likely, and 8% rather unlikely, and similarly for rates of interest below 7%. A particular transaction might actually show a rate of 7.01%, and in fact it is unlikely that any transaction would show a rate of return of exactly that which we might have used in a valuation. In other words, rates of interest derived from the analysis of transactions of a particular class of property can vary continuously over a range, but tend towards a mean. If the distribution of a particular set of data is normal then about 95% of all observations will lie within two standard deviations. Thus if the mean of the expected rates of interest is 7% and the standard deviation is .25%, 95% of possibilities will lie between 6.5% and 7.5%.

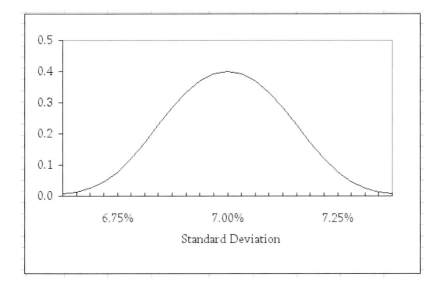

11.2 Simulation

The principle of simulation is an extension of the Scenario approach, using many more trials than can be considered using that facility. Even so, it would be impossible to perform every possible combination of variables. Consider how many valuations would be required if we tried every rate of interest between 6% and 8% with steps of .001% and price per m^2 between £100 and £120 with steps of 1p.

Fortunately a very good approximation can be obtained by taking a sufficiently large random sample in which each of the inputs to the valuation is allowed to vary randomly within defined limits. We would not of course use a random approach for some variables, for example floor areas or the number of rooms which can be ascertained with substantial precision.

11.3 The computer

The computer of course is ideal for this purpose and we can easily set it up to select a random number between two defined limits with the **Randbetween** function. Suppose we are looking for a random price per m^2 between £140 and £160. Select a cell on the worksheet and go to **Insert Menu⇒Function** and select **Math & Trig** and **Randbetween**. Note that this function returns integers only – not decimal fraction parts of numbers.

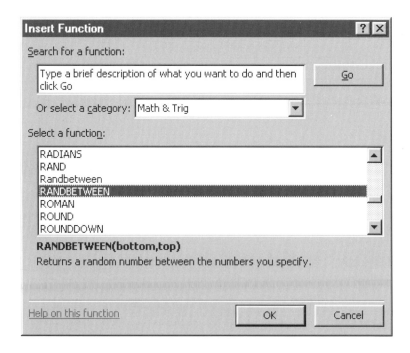

Click on OK and you will be presented with another dialogue box asking for the bottom and top numbers of the range. Insert these and click OK.

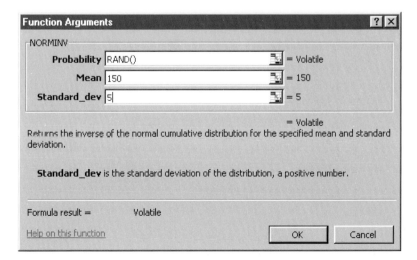

A random number will appear in the selected cell. However, each time the worksheet is recalculated for any reason a new random number between 140 and 160 will appear. You can force the sheet to recalculate by pressing the function key F9. Try it!

This procedure however will not allow for emphasis to be given to the mid-part of the range of numbers. If we want our random prices to be distributed around the mean tending towards a normal distribution we must use the **Norminv** function. Select this function in the same way as before and the following dialogue box will appear.

We want a random number between 140 and 160, so we enter **Rand**() as the Probability and 150 as the mean. Recall that 95% of random numbers fall within two standard deviations of the mean, and therefore if we select 5 as the standard deviation 95% of randomly selected numbers will fall between 140 and 160. A few (5%) will fall outside these limits, but the majority of the numbers will be close to the mean of 150.

11.4 A simple residual appraisal

We will now apply these techniques to a simple residual problem where the site price is fixed but the net development value and construction costs are uncertain. In this over-simplified example an estimation of other costs to the scheme are calculated as 20% of the total construction costs. The following is our starting point.

Appraisal

Net Development Value	£2m
Construction Costs	£1m
Site	£0.5m
Other costs	£0.2m
Profit	£0.3m
Profit on Cost Ratio	15.00%

We first consider the range of possible values for each variable. We will assume that £2m is our estimate of the likely net development value, but that it could possibly vary from £1.7m in a worst possible case to £2.3m in a best possible case. Similarly, we expect the total construction cost to be about £1m but it could vary between £0.9m and £1.1m, and other costs will of course vary with this.

We enter details into the worksheet as follows. Selecting **Tools Menu⇒Options** and clicking **View** and **Formulas** will show the following:

	A	B
1	Appraisal	
2		
3	Net Development Value	=NORMINV(RAND(),2000000,300000)
4	Construction Costs	=NORMINV(RAND(),1000000,100000)
5	Site Cost	500000
6	Other costs	=B4*0.2
7	Profit	=B3-SUM(B4:B6)
8	Profit on cost ratio	=B7/B3
9		

Reverting to normal view, one example of the result is :

	A	B
1	Appraisal	
2		
3	Net Development Value	1,933,272
4	Construction Costs	969,526
5	Site Cost	500,000
6	Other costs	193,905
7	Profit	269,841
8	Profit on cost ratio	13.96%
9		

Note that if you set up this calculation on your computer you will almost certainly get a different answer, and another answer within the set parameters will be calculated if you press F9 or do any other operation which forces the worksheet to recalculate.

11.5 Using a Macro

Of course we could now do a series of appraisals using these randomised inputs and write down the answer to each, total them, and then find the mean and standard deviation, and finally plot a graph of the results to find the output distribution. Obviously this is where we should again be using the computer to do the job. We need to set up a Macro in Visual Basic to perform the repetition.

As we discussed in Chapter 8, a Macro is a routine which carries out a series of operations automatically when called. As before, we run the procedure with the macro recorder turned on and having finished turn it off. We will then need to tweak some of the steps in this recording to make our Macro more efficient by modifying the VB code.

The operations will be as follows :

(a) Having set up the valuation as before, turn on the Macro Recorder with **Tools Menu⇒Macro⇒Record New Macro** ... and in the dialogue box enter the name of this Macro and a shortcut key. Enter 'm' as a shortcut key, disregard the other boxes and click 'OK'.

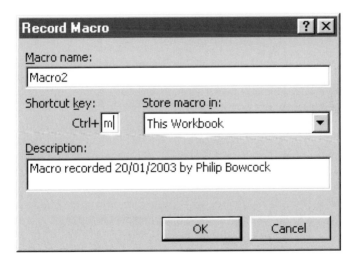

(b) The **Stop** Toolbar appears. From now on, until we click the Stop button, every action will be recorded. To stop go to **Tools Menu⇒Macro⇒Record New Macro** and click **Stop**.
(c) Press F9 to force the computer to recalculate the worksheet.
(d) Select cell B8, and **Edit Menu⇒Copy**.
(e) Select cell D3, click **Edit⇒Paste Special** and **Values**, and then click the stop button. Don't use the **Paste** option as this will paste the cell references in B8 as well and create an error.

We now have a simple Macro, which recalculates the appraisal, and copies the answer to a different cell. To view the code select **Macros⇒Macros⇒Simulate ⇒Edit**.

```
Sub Simulate()
'
' Simulate Macro
' Macro recorded 30/12/2002 by Philip Bowcock
'
' Keyboard Shortcut: Ctrl+m
'
Calculate
Range('B8').Select
Selection.Copy
Range('D3').Select
Selection.PasteSpecial Paste:=xlValues, Operation:=xlNone,
SkipBlanks:= _
False, Transpose:=False
End Sub
```

11.6 Editing the Macro

We pointed out in Chapter 8 (Visual Basic) that the only lines of actual code are those that do not begin with an apostrophe.

The SELECTION.PASTESPECIAL Command has several different options. We selected just one: Values. However Visual Basic records the state of these other options too. Since we require the default settings we can delete all options apart from Values. Our Macro should run in exactly the same way, although it ought to be fractionally (but imperceptibly) faster since it will have fewer lines of code to read.

The Macro is currently only capable of recalculating the appraisal and dumping the profit/cost ratio in one location. Every time we run the Macro the most recent profit/cost ratio in B10 is over-written by the newly recalculated value. The point of the exercise is to record every recalculation, so that we can make an assessment of the behavioural characteristics of this ratio. We need to find a way of getting the Macro to dump each newly recalculated answer in a different cell each time.

At the moment the following line of code in our Simulate Macro is very strict about where the PASTE.SPECIAL operation should take place.

Range('D3').Select

We can ask Visual Basic to run through the code several times automatically, and change this line of code each time it does so. First of all we need to change this line of code to the following.

Range('D3').Offset(i,0).Select

The Offset (i,0) method chooses a cell i rows and 0 columns from B10, for each run through the code. i depends on the number of runs already completed. To tell Visual Basic to run through the code several times we add two other lines to the top and bottom of the code. Finally we add the following to the end to tell the selection to return to D3.

It now reads:

```
Sub Simulate( )
'
' Simulate Macro
' Macro recorded 30/12/2002 by Philip Bowcock
'
  ' Keyboard Shortcut: Ctrl+m
'
For i = 1 To 10
Calculate
Range('B8').Select
Selection.Copy
Range('D3').Offset(i, 0).Select
Selection.PasteSpecial Paste:=xlValues
Next
Range('D3').Select
End Sub
```

The first line of code tells Visual Basic to run through the code setting i as 1 the first time the loop runs. The last line of code 'Next' tells Visual Basic to go back to the top of the code where it will increment i to 2, and continue incrementing each time through the loop until i equals 10. The final selected cell will be D3. Press **Escape** if B8 is still set for copying.

Finally, to make the process run faster, particularly we increase the total number of runs, enter the following at the start and end of the code:

```
Application.ScreenUpdating = False
Application.ScreenUpdating = True
```

Run the edited code and watch Excel produce 10 recalculated results in 10 different cells.

11.7 Interpreting the results

Ten results will not provide any meaningful conclusions. It would be more realistic to change i in the above code by replacing 10 with 1000. However for this example we will edit the code to produce just 100 results.

The statistical interpretation of a set of numbers is vast, but Excel will quickly summarise the most commonly used measures.

(a)   Select **Tools Menu**⇒**Macro**⇒**Macros** ... Select the Simulate Macro in the dialog box, and then **Edit**.
(b)   Substitute the line of code: For i = 1 To 10, with For i = 1 To 100.
(c)   Exit the code, and run the macro again to produce 100 results.

It will be helpful if at this point we name cells D4:D103 as 'Results'.

(d) Select **Tools Menu**⇒**Data Analysis**⇒**Descriptive Statistics**. The following window should appear. (If this Add-in is not present see Appendix B.)

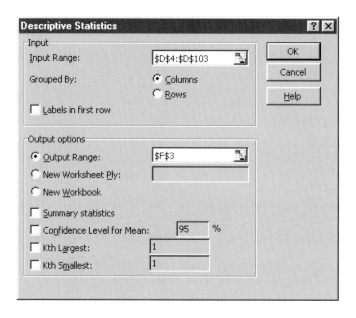

(e) Select the Input Range textbox and enter 'Results'.
(f) Make sure the Columns radio button is selected, and that 'Labels in First Row' remains un-selected.
(g) Select the Output Range radio button so that the results will be dumped in the same sheet. Enter F3 as this range.
(h) Finally click the Summary statistics box, and OK.

You may need to widen the columns so that you can see all the information, but you should see the following table of results.

F	G
Column1	
Mean	306,706
Standard Error	36,894
Median	315,636
Mode	#N/A
Standard Deviation	368,940
Sample Variance	136,116,356,880
Kurtosis	-0.04
Skewness	0.24
Range	1,872,255
Minimum	-530,782
Maximum	1,341,473
Sum	30,670,620
Count	100

The figures produced by your own simulation should be similar but are most unlikely to be exactly the same.

The most useful statistics from this table are discussed elsewhere, although a full discussion is beyond the scope of this book.

11.8 Histogram

The most appropriate graphical representation of our simulation is a histogram. This groups our output into classes, a series of sub-ranges within the total range, from which we can plot the results on a bar chart. Excel can produce a histogram automatically from the output data. The only other items information required are the classes, or bin ranges.

Examine the maximum and minimum values and select convenient bin range values above and below these allowing for dividing the interval conveniently. We will use nine bins starting at –£400,000 with intervals of £200,000 as shown. (The 'Bin Range' – the computer equivalent of a set of plastic boxes into which you sort results.) Enter these numbers into F20:F28.

Now that bin ranges are defined we can initiate the histogram function.

(a) Select **Tools Menu**⇒**Data Analysis**⇒**Descriptive Statistics**⇒**Histogram**. The following window should appear.

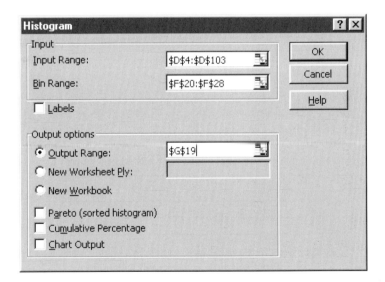

(b) Select the Input Range text box, and then highlight the output column from the simulation or enter 'results'.
(c) In the Bin Range text box highlight the cells F20:F28.
(d) Select the Output Range radio button, ensure that the cursor is flashing in the Output Ranges text box then select cell G19.
(e) Finally tick the Chart Output text box, then OK.

This should provide you with a histogram similar to the one below. Notice that the chart indicates that there is a greater frequency of results on the left hand side (lower returns), than there are on the right (higher returns) . The chart is skewed to the left. We might have concluded this from the table since a value of less than 0 indicates a negative skewness. In the example above skewness has a value of -0.69. Equally, kurtosis is a measure of how flat the histogram is, or how evenly spread the results are. In the above example most of the returns cluster around one or two results making the histogram relatively peaked. Again we might have deduced this from our descriptive statistics. Since kurtosis of less than 3 indicates a more peaked histogram, we are not surprised to learn that the value is in fact 1.53.

Since the numbers to be shown on the x-axis are quite large it will be convenient to divide them by 1,000. Use cells I20:I28 to do this. We should now have something like the following:

Bin	Frequency
-400,000	2
-200,000	7
0	14
200,000	14
400,000	29
600,000	14
800,000	9
1,000,000	8
1,200,000	2
More	1

The final step is to create the histogram chart using cells I20:I28 for the x-axis and H20:H28 for the y-axis. Some minor adjustments to the formatting will produce the chart. Note that the numbers on the x-axis are the upper limits of the particular bin, for example the third bin ('200') show results between 0 and 200,000.

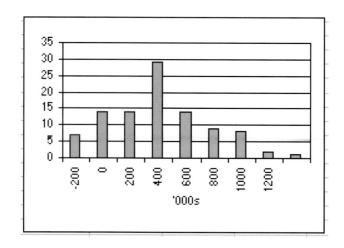

We now have a graphical representation of the distribution of 100 valuations using our randomised inputs. As pointed out earlier, it would probably be very desirable to run this procedure many more times, but extending this to 1000 or 5000 is merely a matter of changing one number in the macro and adjusting references to the results.

11.9 Further consideration of randomness

In the above discussion we have assumed that some of the input data will vary within a normal distribution. However this is not necessarily the most appropriate assumption. Some research has indicated that a skewed distribution may be more appropriate.

For example, we might be using a long-term rate of interest based on Minimum Lending Rate as at January 2002. It would probably be realistic to assume that it will not fall much further but that there is a possibility in years to come that it could rise to levels experienced in the 1980s – in other words that the distribution should be skewed. However incorporating an inverse skewed function into a simulation would require considerable research or experience.

11.10 Conclusion – Negligence

Obviously this procedure was based on very simple valuation data, but the technique can easily be adapted to much more complicated valuations where more of the inputs may be subject to variation.

No doubt every surveyor carries the spectre of a negligence claim in mind in every commission undertaken. Many cases which have come before the courts have accepted that a margin of error of 10% in opinion evidence is acceptable, but in more recent cases the question of methodology has been raised. It is therefore open for consideration that an opinion of value based on a simulation and including the statistical results discussed above may carry greater weight than a single valuation or even a series of scenarios.

Chapter 12

Portfolio management

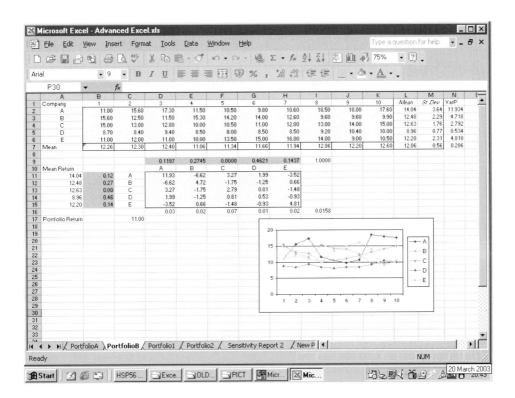

12.1 Portfolio management

Why buy shares in more than one company? For the simple reason that profits from one company can shield the loss from another. If they do well, fine, but they are unlikely to all lose money at the same time, thus significantly reducing the variability, ie risk, of returns.

The management of a portfolio, involving as is usually does, large amounts of money, is in itself a risk-prone operation. In this chapter we set out to show how Excel can be used for this purpose. As always our aim is to demonstrate the operation of the software and not to advocate any particular approach to management problems.

12.1.1 Risk and uncertainty

'Risk' and 'uncertainty' are terms of some imprecision, depending on the context and the user. For our discussion we will use the term 'risk' to cover both concepts, and divide risk into the two categories often referred to as 'systematic risk' (or 'market risk') and 'unsystematic risk' (or 'specific risk').

The difference between these concepts generally is that the former risks are those to which all investment is subject and which cannot be minimized by diversification (extreme examples could include earthquakes, volcanic eruptions, large meteorites and major wars).

Shares may go up or down in value. If they are traded frequently these changes may happen in a fraction of second (and at the time of writing the markets are doing just that as they come to terms with corporate fiascos). Many professionals are dedicated to assessing not just the direction but also the potential size of these changes.

The latter include risks which can be reduced by careful diversification of investments such that under-performance of return on one investment is compensated by over-performance of another. Returns on investments are never fixed or certain. They will depend on the health of the individual company, the market, and the economy as a whole. So far as we are concerned, we shall be considering risk to mean the variability in returns from different investments.

Investment by individuals in property is usually by means of the purchase of shares in property companies, and this is the form which we shall mainly consider. However we should not forget that the property companies themselves are investors, and will be assessing the risks involved in the purchase and/or re-development of individual properties in the same way.

12.2 The statistics

We have already reviewed the Mean and the Standard Deviation which give us a measure of the average and the variation around this in relation to Simulation. We now consider two further statistics that are used to assess these uncertain variations and hence risk. These are two related though distinct statistical measures – Correlation and Covariance. As concepts they may be compared with the Mean and the Standard Deviation of a single variable and their function is to compare the relative trends and variances of two or more variables. We first examine Correlation and its application to our portfolio problem and later the use of Covariance.

12.2.1 Correlation

The Correlation Coefficient is a measure of the relationship between two variables. In the next example we compare the movements of two investments over a period of time to test the extent to which they are related. For example a property company specializing in large shopping centres (of which there are only a few) may purchase a major centre. This may be positively reflected in the share price of that company, but equally the purchase may negatively affect the share price of other property companies who were trying to win the same business. There is then the possibility that as the shares of one property company move up, the others in its sector move down. They are said to be negatively correlated.

Example: Returns over the last 10 years[1] from an investment of £100.00 in each of two property companies have been have been as follows (pounds):

	A	B	C	D	E	F	G	H	I	J	K	L	M
1	Company	1	2	3	4	5	6	7	8	9	10	Mean	Variance
2	A	10.00	12.60	13.30	8.60	8.20	17.00	8.20	10.00	13.00	9.00	10.99	7.529
3	B	10.00	9.50	9.00	10.60	14.20	8.40	12.60	9.20	9.60	12.00	10.51	3.077

This data can be shown graphically:

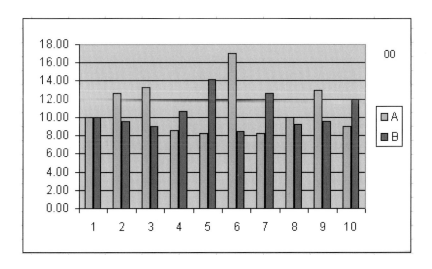

It is apparent from observation of this chart that as returns from one investment have risen in some years returns from the other have tended to fall, and conversely. However before considering this we examine the variance of each investment individually. Both companies showed a marked rise-fall-rise-fall pattern, and these can be measured by calculating the variance of each with the function

=VARP(data_rangeA)

---

[1] For the purposes of these examples we use a time unit of one year. For practical purposes this might be considered too long, but a unit of one quarter or even one month could equally well be used.

The entry for Company A would be :

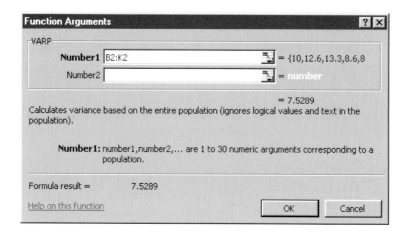

Repeating the process using the second row of data we have 7.529 for Company A and 3.187 for Company B, and these are shown in the end column of the schedule above.

12.2.2 Correlation matrix

To establish the extent to which the stocks co-vary we can determine the correlation coefficient. This statistic varies between 1 and –1. A correlation of 1 denotes perfect correlation – returns always move in the same direction, while a correlation of –1 indicates perfect negative correlation ie they always move in exactly opposite directions. A correlation of 0 means there is no relationship whatever. To compare the extent to which the returns from the two companies are related we use the correlation function using **Insert Menu⇒Function** and selecting **Correl**. Insert the data range for each investment.

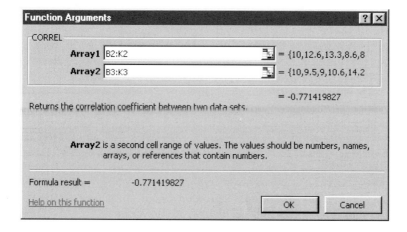

Applying the calculation to these companies we have:

$$=CORREL(B2:K2,C2:K2)$$
$$= -0.6025$$

The value of the correlation gives us a measure of this relationship and the sign indicates that the two stocks are negatively correlated which confirms our subjective opinion of a negative correlation between the two stocks – as the return from one stock increases the return from the other has a strong tendency to decrease.

12.2.3 Covariance

Among the statistical functions available from **Insert Menu⇒Functions** ... are three for calculating the variance[1] of a set of data. VARA is not relevant for our purposes, but it is important to appreciate the difference between the other two, VAR and VARP. The former estimates the variation of a population from the sample we are using (for example estimating the variation in house prices in a town from information provided by a single agent). VARP on the other hand treats the data as the entire population (using the data from all agents in the town). Another function which we shall use is COVAR, a comparison between the variances of two sets of data which treats these as the entire population. It is therefore appropriate to use VARP to consider the variance of a single set of data.

The Covariance function provides a measure of how the variance of one data range is related to that of another data range.

Regarding our example now as a portfolio of the two investments we can consider the variability of each asset in relation to the whole. Covariance compares the variability of the two returns over the period, and in Excel this is calculated by the function:

$$=COVAR(data\_rangeA,data\_rangeB)$$

```
Function Arguments                                        [?][X]
 COVAR
         Array1  B2:K2          [⬚] = {10,12.6,13.3,8.6,8
         Array2  B3:K3|         [⬚] = {10,9.5,9,10.6,14.2

                                      = -3.7129
 Returns covariance, the average of the products of deviations for each data point pair in two
 data sets.

           Array2 is the second cell range of integers and must be numbers, arrays, or
                  references that contain numbers.

 Formula result =            -3.7129
 Help on this function                        [   OK   ]    [ Cancel ]
```

---

[1] The Variance is the square of the Standard Deviation.

which in our example gives –3.505.

Note that the magnitude of the covariance is not limited to the range –1:+1.

12.2.4 Constructing a portfolio of two investments

When selecting investments for a portfolio the selection criteria are extended. It is no longer efficient to select on the basis of the highest return or the lowest risk. A combination of stocks creates a combined return with a new level of risk, and our object is to find the optimum combination to select.

In the case of investment A we found that over the ten years it had produced an average return of 10.99% with a variance of 7.529 %. Investment B on the other hand showed an average of 10.07% and a variance of 3.181%. If we were investing in only one of these our decision would depend very much on our appetite for risk, but there is another option. By investing 60% of our money in A and 40% in B we would have had a new data range with a new average and variance, as shown in Row 4.

	A	B	C	D	E	F	G	H	I	J	K	L	M
1	Company	1	2	3	4	5	6	7	8	9	10	Mean	Variance
2	A	10.00	12.60	13.30	8.60	8.20	17.00	8.20	10.00	13.00	9.00	10.99	7.529
3	B	10.00	9.50	9.00	10.60	14.20	8.40	12.60	9.20	9.60	12.00	10.51	3.077
4	A40 : B 60	10.00	11.36	11.58	9.40	10.60	13.56	9.96	9.68	11.64	10.20	10.80	1.421

For the 60:40 portfolio the average return of would have been 10.44%. The risk of the portfolio however depends on the variance of each stock and the covariance between each pair and to calculate this we set out the following Variance/Covariance matrix.

	A	B
A	=VARP(data_rangeA)	=COVAR(data_rangeA, data_rangeB)
B	=COVAR(data_rangeA, data_rangeB)	=VARP(data_rangeB)

This produces the following results:

	A (40%)	B (60%)
A (40%)	7.529	–3.713
B (60%)	–3.713	3.077

To find the variance of the portfolio we now apply the combined weights to each value in the matrix, as follows:

0.6 * 0.6 *	7.529	1.205
0.4 * 0.6 *	−3.713	−0.891
0.6 * 0.4 *	−3.713	−0.891
0.4 * 0.4 *	3.077	1.108
		0.530

As a result of combining these two stocks in our portfolio we have changed the return to 10.80% – slightly less than A but the risk is reduced to 0.531, which is much less than either A or B.

For two investments the calculation is relatively straightforward, and we could continue experimenting with different combinations to improve the result. For more investments however the calculations become quite unwieldy. The practical way to deal with the portfolio's risk characteristic is to set up a covariance matrix. We now demonstrate how to deal with this otherwise impractical calculation problem using Excel.

12.3 A larger portfolio

Example: In this example we consider five investments available for our portfolio which have the following histories of returns over the last 10 years. As before we assume a total investment of £100 in each at the start of the period, we now wish to invest in a selection so as to minimise risk.

Company	1	2	3	4	5	6	7	8	9	10
A	11.0%	15.6%	17.3%	11.5%	10.5%	9.8%	10.6%	18.5%	18.0%	17.6%
B	15.6%	12.5%	11.5%	15.3%	14.2%	14.0%	12.6%	9.6%	9.6%	9.9%
C	15.0%	13.0%	12.8%	10.0%	10.5%	11.0%	12.0%	13.0%	14.0%	15.0%
D	8.7%	8.4%	9.4%	8.5%	8.0%	8.5%	8.5%	9.2%	10.4%	10.0%
E	11.0%	12.0%	11.0%	10.0%	13.5%	15.0%	16.0%	14.0%	9.0%	10.5%

As a start it may be helpful to consider a graphical representation of these returns:

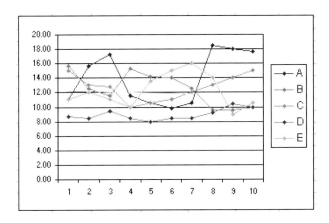

We can see at once that D has produced low returns, but they have not been subject to much variation. Conversely the others have varied considerably over the years. The problem therefore is – what combination of investments would have produced the best outcome?

Before we decide on weighting we set up a correlation matrix as before, entering formulae as follows. Note that this is a typical symmetric matrix with variances on the main diagonal and covariances on the upper triangle. The lower triangle is a reflection of the upper triangle, and these cells can conveniently be set equal to the respective cells above.

The first step is to enter the data into the worksheet:

	A	B	C	D	E	F	G	H	I	J	K	L	M	N
1	Company	1	2	3	4	5	6	7	8	9	10	Mean	St. Dev	VarP
2	A	11.00	15.60	17.30	11.50	10.50	9.80	10.60	18.50	18.00	17.60	14.04	3.64	11.934
3	B	15.60	12.50	11.50	15.30	14.20	14.00	12.60	9.60	9.60	9.90	12.48	2.29	4.718
4	C	15.00	13.00	12.80	10.00	10.50	11.00	12.00	13.00	14.00	15.00	12.63	1.76	2.792
5	D	8.70	8.40	9.40	8.50	8.00	8.50	8.50	9.20	10.40	10.00	8.96	0.77	0.534
6	E	11.00	12.00	11.00	10.00	13.50	15.00	16.00	14.00	9.00	10.50	12.20	2.31	4.810
7	Mean	12.26	12.30	12.40	11.06	11.34	11.66	11.94	12.86	12.20	12.60	12.06	0.56	0.286

We set out the formulae for the covariances. Below are entries for the cells in two of the matrix columns. Follow the pattern for the other cells, using the **Copy** function where possible rather than re-typing every cell. Set the upper triangle cells equal to the corresponding lower triangle cells.

	A	B	C	D	E	F
8						
9				0.2	0.2	0.2
10	Mean Return			A	B	C
11	=L2	=D9	A	=VARP(B2:K2)	=D12	=D13
12	=L3	=E9	B	=COVAR(B2:K2,B3:K3)	=VARP(B3:K3)	=E13
13	=L4	=F9	C	=COVAR(B2:K2,B4:K4)	=COVAR(B3:K3,B4:K4)	=VARP(B4:K4)
14	=L5	=G9	D	=COVAR(B2:K2,B5:K5)	=COVAR(B3:K3,B5:K5)	=COVAR(B4:K4,B5:K5)
15	=L6	=H9	E	=COVAR(B2:K2,B6:K6)	=COVAR(B3:K3,B6:K6)	=COVAR(B4:K4,B6:K6)
16				=SUMPRODUCT($B$11	=SUMPRODUCT($B$11	=SUMPRODUCT($B$11

Results now appear as follows:

	A	B	C	D	E	F	G	H	I
8									
9				0.2000	0.2000	0.2000	0.2000	0.2000	1.0000
10	Mean Return			A	B	C	D	E	
11	14.04	0.20	A	11.93	-6.62	3.27	1.99	-3.52	
12	12.48	0.20	B	-6.62	4.72	-1.75	-1.25	0.66	
13	12.63	0.20	C	3.27	-1.75	2.79	0.91	1.10	
14	8.96	0.20	D	1.99	-1.25	0.81	0.53	-0.93	
15	12.20	0.20	E	-3.52	0.66	-1.48	-0.93	4.81	
16				1.41	-0.85	0.73	0.23	-0.09	0.2857

We can examine the return and risk for each individual stock before determining a portfolio by applying the **Average** and **Standard Deviation** functions to each data range, as in Columns L and M above.

The return for this portfolio is calculated as before and gives an average of 12.06. The portfolio risk depends on the correlation between the individual stocks. Using the covariance matrix we sum the weighted covariances.

We can do this one of two ways.

METHOD 1, using the **Sumproduct** function.

(a) Add the weightings to the matrix, as above.
(b) Below the first column in the matrix enter the function.
(c) SUMPRODUCT($B$11:$B$15,C3:C7) to give 2.10.
(d) Copy this across through to cell H16.
(e) Finally in cell I16 enter the function

$$SUMPRODUCT(D9:H9,D16:H16).$$

This generates this riskiness of the new portfolio of 0.7023.

METHOD 2. Using the MMULT function shortens the calculation by calculating the bottom row of results in one go. All we need in this instance is the portfolio weightings listed horizontally as they are currently in the top row.

Select cell I16 and enter this function.

$$=SUMPRODUCT(MMULT(C1:G1,C3:G7),C1:G1)$$

to give the same result.

So we have an expected return of 12.06 and a risk of 0.7023 which compares favourably with our analysis of the individual stocks above. The above used equal weighting, but we could keep experimenting with different weightings to try to improve on the risk and return combination. For instance an equal share of 20% in each stock would produce an expected return of 12.06 and a risk of 0.7023. What we would like to know is which combination produces the highest return for the lowest amount of risk. Instead of trial and error we can use Excel's Solver function.

12.4 An application of Solver

We considered the Solver function in Chapter 10 in connection with a development scheme, but the same technique can be used to find the optimum combination of investments to give the minimum risk. Remember that risk is a trade-off against return and vice-versa.

We start by asking the question: 'What combination of weights/proportions will give us a specified level of return with the lowest risk?' In this case five variables – the proportions of each investment – need to be considered, and in addition there are two constraints on what values the variables can take. All the weights must be positive and the total must add up to 1 otherwise the proportions would be meaningless. We assume that we require a return of 11.0%.

To find the optimal proportions initiate the Solver function by selecting **Tools Menu⇒Solver** as we discussed in Chapter 10. Our starting point for the data is the covariance matrix which we derived before. We also assume as a starting point for this operation that the same sum – 20% of the total – was invested in each. The total return is then the product of the mean returns and the proportions:

	A	B	C	D	E	F	G	H	I
8									
9				0.2000	0.2000	0.2000	0.2000	0.2000	1.0000
10	Mean Return			A	B	C	D	E	
11	14.04	0.20	A	11.93	-6.62	3.27	1.99	-3.52	
12	12.48	0.20	B	-6.62	4.72	-1.75	-1.25	0.66	
13	12.63	0.20	C	3.27	-1.75	2.79	0.81	-1.48	
14	8.96	0.20	D	1.99	-1.25	0.81	0.53	-0.93	
15	12.20	0.20	E	-3.52	0.66	-1.48	-0.93	4.81	
16				1.41	-0.85	0.73	0.23	-0.09	0.2857
17	Portfolio Return		12.06						
18									

Note the following:

(a) The proportions of funds to be placed in each investment appear above the investment label (D9:H9).
(b) Cell I19 is the sum of D9:D9, and must equal 1.00.
(c) Cells B11:B15 are set equal to D9:H9 respectively.
(d) Cell I16 is the risk of the portfolio and this is the value to be minimised.
(e) The return of this portfolio appears in cell C18.

The Solver Parameters window should look like this when the constraints have been entered as described in Chapter 10, and it then remains just to click on the Solve button to complete the operation.

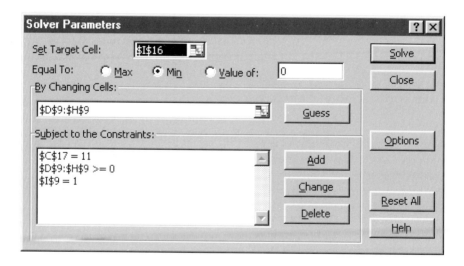

The final solution appears, and we can see immediately that by changing our investments to the proportions shown in B11:B15 the return is now 11.0% as required and the risk is 0.0158, compared with the previous risk of 0.2851.

	A	B	C	D	E	F	G	H	I
8									
9				0.1197	0.2745	0.0000	0.4621	0.1437	1.0000
10	Mean Return			A	B	C	D	E	
11	14.04	0.12	A	11.93	-6.62	3.27	1.99	-3.52	
12	12.48	0.27	B	-6.62	4.72	-1.75	-1.25	0.66	
13	12.63	0.00	C	3.27	-1.75	2.79	0.81	-1.48	
14	8.96	0.46	D	1.99	-1.25	0.81	0.53	-0.93	
15	12.20	0.14	E	-3.52	0.66	-1.48	-0.93	4.81	
16				0.03	0.02	0.07	0.01	0.02	0.0158
17	Portfolio Return		11.00						
18									

The alternative approach would be to specify a degree of acceptable risk by setting Solver as follows:

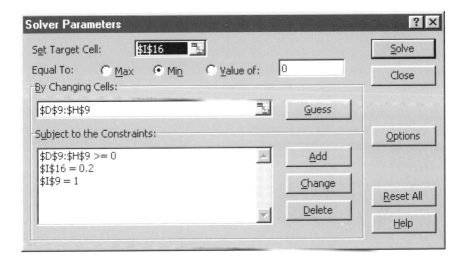

The return with this level of risk is 12.61%, and one can see from this the trade-off between risk and return, and the sensitivity involved.

It is essential to appreciate that this is an analysis of past events, and we have considered returns over the previous 10 years. This is a long time in finance, but of course it could equally have been a half-yearly, quarterly or monthly analysis which would accord more with commercial practice.

You should also appreciate that it is not always possible for Solver to find a solution, and this applies to any linear programming problem. If we change the required return in the example above to 25% there is clearly no way in which this can be achieved whatever the risk.

The main problem of course for advisors is to consider future expectations of yields but analysis of the past is valuable when making decisions about prospective investments. Unfortunately our clairvoyance skills do not enable us to give substantial advice on future events – that we leave to you!

Appendix A
Bibliography

Financial Mathematics

Real Estate Investment. A Capital Market Approach, G. Brown and G. Matysiak, Prentice Hall 2000

Principles of Corporate Finance, R. Brearley, S. Myers, McGraw-Hill 1996 chapters 2–3, 5

Property Investment Appraisal, A.E. Baum, N. Crosby, International Thompson Business Press 1995

Practical Application of Worksheets

Excel for Surveyors, Philip Bowcock, Natalie Bayfield, Estates Gazette 2000

Introduction to Valuation, D. Richmond, Macmillan 1994, Computer Applications pp. 63–173

Property Development Appraisal & Finance, D. Isaac, Macmillan 1996 pp. 111–114

Property Valuation Techniques, D. Isaac, T Steley, Macmillan 2000. Computer models and the use of worksheets pp. 175–188

Worksheets and valuation. Mainly for Students, Estates Gazette 21–1–1995

Microcomputers in Valuations. Mainly for Students, Estates Gazette 11–1–1996

Macros

Excel 2002 VBA, Bullen, Green Bovey Rosenberg, Wrox 2001

Appendix B

Add-Ins

Excel is a very complex and powerful application, and has so many facilities that it is most unlikely that anyone would ever use all of them. In particular there are several options for data analysis which are fairly specialist, and in order to cut down on complexity for those who might never use them, they are available as 'Add-Ins' which can be called if required.

At the time of original installation of the Office suite on your computer you have the option of installing or otherwise, dozens (in the latest XP version of Office) of optional facilities including the Add-ins for Excel. If these were not installed at the time you (or your Administrator) will need to insert the original disk and run the installation procedure again. (It might be that these optional extras were not installed because of shortage of disk space at the time.)

Assuming that the Add-Ins have been installed we now need to attach those we need to Excel. To do this, go to **Tools Menu⇒Add-Ins** and, for the purposes of the examples in the foregoing chapters, click on the **Analysis Toolpak** and **Solver Add-In** boxes and OK. You will now find that these have been added to the **Tools Menu**.

Appendix C

A note on Excel for Macintosh

Most of the facilities and examples which we have discussed in previous chapters apply equally to Excel for Macintosh, but there are a few differences – some additional features and some which work slightly differently, and some commands are found in different menus. However most Macintosh users will be able to find these differences and additional facilities. Files written on PC can be read by Macintosh and vice versa without any modification except that a file written on Macintosh should have the .xls suffix added to the name so that a PC can recognise it as a worksheet.